The Infancy
of Particle Accelerators

Life and Work of Rolf Wideröe

Compiled and Edited
by Pedro Waloschek

Rolf Wideröe

The Infancy of Particle Accelerators

Life and Work of Rolf Wideröe

Compiled and Edited by Pedro Waloschek

vieweg

Title of the German original edition: 'Als die Teilchen laufen lernten –
Leben und Werk des Rolf Widerøe'
This English translation is an updated and improved version of the
German original edition.
Translated by Karen Waloschek, London.

Typesetting and layout by Pedro Waloschek

Printed on acid-free paper

ISBN 978-3-663-05246-3   ISBN 978-3-663-05244-9 (eBook)
DOI 10.1007/978-3-663-05244-9

Contents

List of Boxes

List of Figures

Introduction

by Pedro Waloschek

The following autobiographical account of Rolf Wideröe's life and work is based on manuscripts and letters written by himself, most of them especially for this report. Data from audio and video recordings with his illustrations and from my notes taken during a series of meetings between the two of us were also included. Rolf Wideröe gave me access to many of his publications and to other documents from which I have extracted further information.

I have compiled, edited and, where necessary, put the texts in chronological order. These were then corrected and supplemented by Rolf Wideröe during the course of several readings. The English translation was also checked by Wideröe and we were able to add some improvements and corrections. This account therefore stands as an authorised biography and is written in the first person. Mrs. Wideröe's accurate memory was of great assistance.

The emphasis has been on Rolf Wideröe's life story and the first developments which led to modern particle accelerators. Technical and scientific comments have been kept as comprehensive and concise as possible. For further details the reader is referred to the many publications quoted in the text and to the extensive literature available, such as the beautiful books 'The Particle Explosion' by Frank Close, Michael Marten and Christine Sutton [Cl87] and 'From X-Rays to Quarks, Modern Physicists and their Discoveries' by Emilio Segrè [Se80], as well as the classic textbook 'Particle Accelerators' by Stan Livingston and John Blewett [Li62] which all contain a great deal of historical information.

An important foundation for this report was provided by Wideröe's extensive notes of a 1983 interview with the two Norwegian physicists Finn Aaserud and Jan Vaagen in Oslo. The article they wrote on him appeared in the magazine 'Naturen'

[Aa83]. Wideröe kept a record of the question and answer session as well as of several observations made by his friends Olav Aspelund and Olav Netteland who were also present (Gunnar Thoresan, the First Curator of the Technical Museum in Oslo was also at the interview). In 1991 Wideröe freely translated all this – and the Naturen article – into German [Wi91], added some comments and partly modified it for use in this report.

The many documents on Rolf Wideröe which are kept in the 'History of Science Collections' of the Library of the 'Swiss Federal Institute of Technology' (ETH) Zurich have been invaluable. After all, he taught at the ETH in Zurich for twenty years. Seventeen volumes comprise all publications and papers as well as Wideröe's patents [Wi70]. Many other documents, such as letters, manuscripts, photographs and audio and video tapes have also been preserved. In future, all other relevant documents regarding Rolf Wideröe's life and work will also be kept there, including those which I have compiled for this account. The Wideröe documentation at the ETH was founded by the head of the 'History of Science Collections', Dr. Beat Glaus and is now maintained and being extended by Mr. Morten Guddal. I am very grateful to both of them, as well as to the archive's staff for their valuable assistance.

I have inserted boxes and a chronological survey which contain some additional points of information. These are generally about interesting parallel developments or events, but also include data with which I aim to assist the reader in obtaining a better general overview. Within my limitations I have tried to verify the data used here by comparing them to various publications and by consulting witnesses of the relevant historical events. In doing so I was able to correct several errors which had crept into accounts of Wideröe's life. I would be most grateful to receive any further corrections or suggestions for improvement, and these should be forwarded to my address: DESY, Notkestr. 85, D - 22603 Hamburg. For any errors which I have added during editing and correcting I beg forbearance and accept full responsibility.

2

While preparing this report I have had the encouragement and active support of many of my colleagues as well as the Directorate of the DESY research centre. Many friends and acquaintances collaborated on my enquiries, among them Dr. Arnold von Arx, Dr. Olav Aspelund [As82], Professor Jean Pierre Blaser, Ing. Heinz Bergmüller, Ing. Derek Darvill, Ing. Christian Falland, Mr. Rüdiger Giel [Gi93], Dr. Thomas Naumann, Mr. Klaus Seib, Dr. Sigmund Nowak, Dr. Jochen Seibert, Ing. Alfred Stüben and Mr. Klaus Thamm. Professor Roald Tangen helped me to clarify some important historical details. Of particular importance were several remarks of the late Professor Wolfgang Paul and some of his publications. Dr. Maria Osietzki [Os87] [Os88] and Mr. Edgar Swinne [Sw92] [Sw93] also supplied me with highly interesting data. I would like to take this opportunity to thank them all.

From the beginning Mr. Wolfgang Schwarz (Vieweg Publishers) supported me in the planning and editing of the book and the publisher's staff gave us excellent support during production. I am very grateful to my daughter Karen for her careful translation and patient updating and to Mrs. Gisela Lüscher, Mr. Derek Darvill and Mr. Russell Bevington for their attentive proof-reading.

I would like to thank Mrs. Gabriele Heessel for transcribing and correcting not only Rolf Wideröe's extensive, hand-written notes, but also many hours of audio recording. And, last but not least, I would like to mention that without the patient assistance and care of my wife Edith this report would never have been completed.

However, before I let Rolf Wideröe speak for himself, I would like to briefly summarize the highlights of both his life and the contributions he made to research and technology; this is especially for those readers who are not yet familiar with his work or, so to speak, as a taster of the account that follows.

It was August 1958 when I first heard of Rolf Wideröe. During a meeting of physicists in Varenna on Lake Como my friend Bruno Touschek told me of a brilliant Norwegian engineer for whom he had worked in 1943. This engineer had brought Touschek schnapps, cigarettes and his beloved books after the Gestapo had imprisoned him in Hamburg's Fuhlsbüttel jail as a result of his fondness for reading foreign magazines. The engineer thought he had had a magnificent idea; he wanted to build a far more effective 'atom smasher' than had ever been possible before. And this in 1943 during the War – in Hamburg. He had applied only well-known laws of physics, and therefore Bruno, as a theoretical physicist, thought that Wideröe's ideas were not publishable at all as scientific work. They seemed to him far too trivial and half-baked. Wideröe however would not let go and submitted his ideas for patenting. This is now regarded as the invention of the 'storage rings' which today are used throughout the world and find their application in fundamental research as well as for many practical purposes.

In Varenna, Bruno Touschek and I continued to speak at length about Wideröe's genius and about the differences between scientific publications on the one hand, and patents, such as are usual in industry, on the other. We also discussed the curious interaction between industrial interests, technical developments, research and politics, especially during the War, which, as we shall soon learn, played an important role in Wideröe's life.

Among experts, Wideröe is generally regarded as the 'grandfather of modern particle accelerators', as the inventor, or co-inventor, of probably the most important ideas on the subject this century, and perhaps even a legitimate candidate for the Nobel Prize. Some of Wideröe's work did not become known among physicists until relatively late; after all, patents do not generally feature in scientists' required reading lists. Many of his ideas were therefore rediscovered by others or had been developed simultaneously. However, this does not in any way affect the historical facts or the value of Wideröe's creative and constructive work. Moreover, Wideröe is an extremely interesting and multitalented person.

Rolf Wideröe was born in Oslo on July 11, 1902. In 1922, that is, when he was twenty, he had already dreamt up the 'ray-transformer', later to become famous as the 'betatron'. This is the theme which runs like a thread through his entire life. He then made drawings and calculations into his notebooks. In 1926, he tried to submit this as his thesis for a doctorate in engineering at Karlsruhe Polytechnic, where it was rejected outright.

Nevertheless, his ideas were understood in Aachen, but his 'ray-transformer' refused to function. Wideröe thus went on to build a 'straight-on' or linear accelerator which did work. Although he only had 25,000 volts at his disposal, with this device he accelerated atomic nuclei as if 50,000 volts were available to him. It was the birth of the 'linac' and the basic principle for the development of all modern particle accelerators. This finally earned Wideröe his doctor's degree in electrical engineering.

In California, Ernest Orlando Lawrence saw Wideröe's thesis published in the magazine 'Archiv für Elektrotechnik', and from the illustrations (he knew very little German) deduced the principle with which he went on to invent the famous cyclotron and for which he was eventually awarded the Nobel Prize. Lawrence always made a point of quoting these facts, and this explains why Wideröe is now so well known in the USA.

Following his dissertation, Wideröe went into industry where he built relays – first in Berlin and then in Oslo. These were probably the best relays available at that time for interrupting the current after short-circuits in power lines. They also indicated the distance from the relay at which the short-circuit happened. The best available relays were later manufactured in Norway and employed in other countries as well. Wideröe did not just develop and build these relays, he also sold them for an electricity company and would even sometimes deliver and install them.

In 1942, hoping to free his brother Viggo, a pioneer of Norwegian aviation and an active participant in the resistance from German imprisonment, Rolf Wideröe agreed to go to Hamburg to build a 'ray-transformer', or 'betatron', which could produce powerful X-rays, following the successful work done at Illinois by Donald Kerst. In any case, this had been a dream of his since youth. A few experts of the German Air Force had thought up the idea of

5

using X-rays against enemy aircraft. However, Wideröe knew nothing of this at first, and serious physicists eventually persuaded the Luftwaffe to drop this plan. However, the Hamburg betatron was successful and ended up as booty of war in England where it served to X-ray large steel slabs. Wideröe on the other hand ended up in a Norwegian prison as a collaborator. The famous scientist Odd Dahl and a few other friends managed to persuade the Norwegian authorities of Wideröe's innocence: he was released after 48 days.

Wideröe was still in Hamburg (1943) when he wrote down his ideas about the 'storage rings' whereby particles, running in opposite directions (stored in circular orbits in vacuum chambers), were to be made to collide. The German patent was kept secret during the war and was retrospectively recognized and published in 1953. In 1956 the same principle was proposed again in the USA by Donald Kerst, Gerry O'Neill and others – without their having had any knowledge of Wideröe's patent. Similar ideas were also proposed in the Soviet Union. In 1961 Bruno Touschek and his colleagues at the Frascati Laboratories near Rome managed to run the first 'storage ring' built according to this principle. In today's high energy physics, storage rings with colliding beams are the main instruments used to investigate the smallest constituents of matter – in essence following Rolf Wideröe's original ideas.

From the very beginning of his accelerator studies (in 1922) Wideröe was concerned about the stability of the orbits of charged particles in rings. In 1945 this concern resulted in a Norwegian patent (submitted in January 1946) which included many formulas and contained the most important ideas required for the construction of a 'synchrotron'. Similar suggestions were being put forward at the same time (1945) in the USA and USSR, by Edwin McMillan and Vladimir Veksler. They led to the construction of the first large circular accelerators.

After the War Wideröe built betatrons for Brown Boveri & Co. (BBC) in Switzerland. Over the years, a total of 78 were delivered and installed. Some of these served to X-ray large industrial components, but most were used in hospitals for radiation therapy on cancer patients. For this reason Wideröe began to dedicate himself to studying the effects of radiation on living cells and on the human body. His proposed theory on this subject, the 'two-components-theory', drew great attention. Wideröe's work in this field was highly influential in instigating the so called 'megavolt-therapy', which utilized high energy electrons and X-rays (of up to 45 MeV) to treat deeply situated tumours. Today it is success-fully applied in thousands of hospitals all around the world – mainly by using small linacs, the descendants of the first one built by Wideröe in Aachen.

Wideröe, who in 1962 was awarded an honorary doctorate in engineering from the 'Rheinisch-Westfälische Technische Hoch-schule' (RWTH) in Aachen, in 1964 received an honorary medical doctorate from Zurich University as well as many other distinc-tions. He was a teaching professor at the ETH in Zurich from 1953 to 1973.

When the first larger particle accelerators were built at the two research centres CERN in Geneva and DESY in Hamburg, Wideröe was called in as a consultant. His advice was always greatly appreciated. Wideröe's consistently interesting questions, com-ments and suggestions can be found in the proceedings of many an international conference on particle accelerators.

Nowadays Rolf Wideröe and his wife Ragnhild live a happy pensioners' life in a lovely house on a hill with a view over the Obersiggethal-Valley and the Swiss city of Baden. Every Saturday he welcomes his children and grandchildren for lunch, and every year he celebrates his birthday with friends and relatives in Oslo. He likes to stop over in Hamburg where he visits old friends, including those at the DESY research institute. It is with astonish-ing freshness and enthusiasm that he recounts his life and work.

Hamburg, March 1994

Fig. 1.1: Rolf and Ragnhild Wideröe in Nussbaumen, October 1992, during a shooting break whilst recording the video 'Wideröe on Wideröe' [Wa83].

Wideröe on Wideröe

1 Family, Youth and Lord Rutherford

If I am going to recount my life, it may be a good idea to start with my family history – although that's not quite as easy as it sounds – and then tell a little about my youth.

Theodor Wideröe, my father, was born the son of a vicar in the Norwegian town of Kongsvinger. He was a businessman, a general agent for French wines and Cognac (Martell) and for Dutch vegetable oils used in the manufacture of margarine. His particular interest lay in postage stamps and he loved the outdoor life. We often went on skiing tours in Nordmarken together and we got on very well. We were a well suited pair.

My grandfather's name was Paulus Peter Marcus Wideröe and he lived between 1827 and 1891. His ancestors can be traced far back. The founding father was Aage Hansen who lived near Molde and also in Veöy, the island which has Odin's 'Ve' relic. In Molde, he married Synnöve Oudensdatter of the famous Aspen family which originally came from Brandenburg and used to be known as Kane. The first historical reference to them dates back to 1340 and they are mentioned again in 1597. This was my father's family.

My mother's forebears originated in Germany and they too have an interesting history. My maternal grandfather was called Carl Gottlieb Launer and was born 1819 in Düro-Brockstadt, south of Breslau. He died in Halden (Norway) in 1902. We suspect that the name Launer came from the Huguenots who emigrated from France during the reign of Frederick the Great.

This grandfather wanted to become a brewer and, as a journeyman, he walked all the way to Constantinople and then back to Vienna where, during an uprising in 1848, he took part in a few battles. He became a captain on the side of the rebels. He had a wife during this period, and after sustaining an injury in one of the battles, she hid him in an oven and nursed him back to health. But then his wife died, so he went back on his travels. He came to Northeim near Hanover where he married Johanne Dorthea Magrethe Cramer, my grandmother. She was born in 1837 in Northeim and died in 1925 in our home in Oslo. She was a white-leather tanner's daughter. The couple moved to Halden (Norway) where he became a master brewer. This is where my mother was born in 1875. She died in 1971 in Oslo.

My grandfather later became a master brewer in Hamburg, but some years later he returned to Halden. It is quite possible that I inherited my wanderlust as well as a few other characteristics from my grandfather.

At this point I would like to recount a story which I think is rather curious. I had four cousins in America, in Seattle, the sons of one of my mother's sisters. During a visit, Orwill Borgersen, the eldest of the brothers, told me of an incident; he was driving around in his car when he accidentally slid into a ditch. A farmer who lived nearby pulled him out and, while doing so, he told my cousin that his father had originally come from Germany, namely Hamburg. While still there, he and his horses had been employed to deliver Master Brewer Launer's beer. This has to be an almost unbelievable coincidence!

I remember that as a twelve or thirteen year old boy, I was already very interested in the natural sciences, particularly physics, and in technology – although I wasn't particularly encouraged in that direction at home. I even built an electric telegraph which connected to a friend who lived next door. My family was a little concerned about some of my chemical experiments. They must have been afraid that I would blow up the house, but it never quite came to that. My two brothers, Viggo (born 1904) and Arild (born 1907) were interested in nothing but aviation, and my sister Else (born 1913) had quite different concerns. My two brothers later founded an airline which was probably the first in Norway. In any case, they set up the first postal link to the north of the country, between Oslo and Stavanger and are therefore regarded as pioneers of Norwegian aviation. Viggo usually acted as the pilot and Arild was the mechanic, although he knew how to fly too. Arild crashed whilst flying over the Oslo Fjord in 1937 and he was killed together with our uncle and aunt. His plane had been brand new, but one of the wings' supports had a bad weld and broke off.

In the 1930s, Viggo had a contract with the shipowner and Antarctic whale fisher Lars Christensen. His task was to take cartographic photographs of the coast and bordering areas of the Antarctic. During one of his reconnaissance flights he discovered a large massif now called 'Sör-Rondane', and one of its mountains was named after him 'Wideröe-fjeld'. It is 3,000 m high. The sections of the Antarctic which were explored on the basis of those reconnaissance flights were subsequently awarded to Norway.

The airline which Viggo and Arild founded still exists and is run in collaboration with SAS and Braathens SAFE. It is known as 'Wideröes Flyveselskap'.

In an article recently published in a Norwegian magazine Viggo was described in very romantic terms: 'He likes to have air under his wings; with his nest way above the city and the fiords, with a broad view over the Bunnefjord up to the Sörkedal-Valley, the sharp eyes above his eagle nose follow the way of the sun, the swallow's flight and the correct arrival time of the WF-782 from Brönnöysund' [Sa93]. Viggo also has a house in Spain and every spring we have a few weeks holiday with him. But normally he lives in Oslo. Needless to say, we get on very well.

In Oslo I had a good friend in Kaare Ström who later became professor of geography and limnology, also in Oslo. His father subscribed to the magazine 'The World of Nature' which I often read when I visited his home, and many articles made an impression on me. For example, in the magazine the splitting of the atom was explained and this interested me greatly. Even then, I had an idea that one could use very strong magnetic fields to force the valence electrons of the atoms onto smaller and smaller orbits, in

11

something like a Super-Zeeman-Effect, and that this may cause the atoms to collapse. Later, it must have been in 1983, I found out during a physicists' meeting in Geilo that it was in fact possible to achieve something like that with magnetic fields of 10^{10} Gauss, and that fields of up to 10^{12} Gauss exist in neutron stars.

While I was at school I wrote to Professor Brock at the University of Oslo and asked him about spectral lines. I received a polite reply with references to books in which I could find out more about my questions. This had been my only contact with the world of physics.

I read many books in those days, such as, Rider Haggard's adventure stories about Africa, Conan Doyle's 'The Lost World' and Övre Richter Frich's books about Jonas Fjeld, as well as many novels serialized in magazines.

But I also found much to interest me at grammar school. The things I learnt there were probably of the greatest use to me later on, and a lot of it must have committed itself to my memory. I was a relatively ordinary student, although private study of the lovely booklets in the 'Göschen Collection' enabled me to learn a few things about higher mathematics. We also had a teacher of mathematics, captain Löken, who was a member of the Norwegian Mathematics Association, so I too became a member of this association. During my last years at school I read something about Einstein's theory of relativity. It must have been around the end of the First World War that the deflection of light by the sun was proven and thus Einstein's theory confirmed. At the age of seventeen I gave a talk on this and on Einstein's theory of relativity. Planck's quanta also interested me. My physics teacher knew nothing about this, so I had to explain it to him.

However, I also studied electromagnetic phenomena, that is, the laws of electrostatics, as well as the laws of induction and their strange equations, which were already being used a lot in technical applications.

In 1919 I was deeply impressed by the news that Rutherford was able to disintegrate the nuclei of nitrogen atoms by bombarding them with fast alpha particles from a radioactive substance (I guess it was radium). I had found out about this through newspapers and magazines. So the alchemists' dream had finally come true!

It was clear to me even then, that natural alpha rays were not really the best tools for this task; many more particles with far higher energy were required to obtain a greater number of nuclear fissions. I thought that perhaps this was a case where solutions could be found with the help of high voltage technology.

I knew that electrically charged particles such as atomic nuclei or electrons could be accelerated by electric fields. The energy thus yielded would correspond precisely to the 'volt-number', which is the voltage-difference traversed by the particles. At a million volts this is a mega-electron-volt or one MeV.

However, it is not possible to increase the voltage indefinitely; very quickly a breakdown happens in the form of a spark or something like a flash of lightning. On a dry day and in a large room it is possible to charge a smooth and sufficiently large metal sphere up to a few million volts. But after that, discharges will happen. In those days this was impressively demonstrated, occasionally even in schools, albeit on a smaller scale.

A further disadvantage of accelerating particles with high voltages is that either the source of the particles or the measuring instruments (or even both) have to be at high voltage, which makes any operation rather awkward and even dangerous.

Furthermore, the maximum of several million volts available to accelerate charged particles which can be achieved with this kind of apparatus is not really all that much, if compared with the energy of alpha rays of radioactive substances; these lie between 5 and 10 MeV which would correspond to an acceleration with 5 to 10 million volts.

Therefore, anyone wanting to achieve such high or even higher particle energies had to look for completely new methods of accelerating particles. And that is where I saw certain possibilities in the elegant, but not easily comprehensible equations of electric-

Box 1

Sir Ernest, Lord Rutherford of Nelson

There has hardly been another scientist this century who has had as much influence on the study of the structure of matter as Lord Ernest Rutherford. As far back as 1908 he received the Nobel Prize for chemistry because he had recognized that radioactive alpha rays were in fact helium particles which were emitted by particular atoms.

In 1911, Rutherford proposed a strange experiment to his then assistant, Hans Geiger (who later developed the Geiger-Müller counter) and to his student Ernest Marsden. He got them to shoot alpha rays at gold atoms. Most of them passed through the gold atoms with practically no hindrance, but a few bounced off, some even backwards.

From this experiment Rutherford deduced that atoms are practically empty, except for a small nucleus in which almost their entire mass is concentrated. This was the discovery of atomic nuclei.

However, of particular interest to Wideröe was the discovery of the nuclear disintegration, which Rutherford had published in the 'Philosophical Magazine' in 1919 after verifying his experimental results for about three years. This found an appropriate echo in the media of the time.

The most important aspect of Rutherford's experiments however, was the method. When nuclear particles collide, it becomes possible to investigate their properties. The main interest in those days lay in researching the composition of atomic nuclei by this method. Nowadays we call this 'scattering experiments'. The higher the energy employed, the smaller are the details of the structures which can be investigated. Moreover, new particles can be generated in this way. This is the method used today to investigate the smallest constituents of matter.

Rutherford's intentions were to find better conditions for his experiments and he encouraged his colleagues to produce particles of higher energy in the laboratory. However, knowledge of this did not reach Wideröe, who was working in Karlsruhe and Aachen, as he had no links with this particular research centre.

Ernest Rutherford, born 1871 in New Zealand, was made a Peer of the Realm in 1931. He died in 1937.

ity and magnetism which already interested me then. They were extensively used in technical fields. That, therefore, is how my desire to study electrical engineering came about. But in any case this subject interested me more than any other.

Then came the decision to attend a German university. My parents were convinced that I would have to go abroad to study in order to fulfil my dreams. They claimed that the Polytechnic in Trondheim, the only one in Norway which ran technological courses, was not suitable for me and even categorised it, rather condescendingly, as a 'kindergarten'. I cannot assess whether it really was like that. I am sure that my parents would have revised their judgement a few years later, but I didn't make many enquiries

Fig. 1.2: Rolf Wideröe, the eighteen year old grammar school boy in Oslo.

15

about this institution at the time. It had only been founded in 1910 and in my time it had about 100 students, as Jan Vaagen later told me during our interview in 1983. The kind of technical training which would have fulfilled my expectations was not available in Oslo where we lived.

However, I was quite happy to go abroad and was particularly interested in Darmstadt and Karlsruhe. I can no longer remember why I chose Karlsruhe in particular. Perhaps the decision was influenced by Professor Richter who was an important figure in the field of electrical engineering in those days. I firmly believed one had to be an academically qualified engineer if one wanted to achieve anything in life.

After sitting for my A-level exams (Examen Artium) in the summer of 1920 at the Halling School in Oslo, my father took me to Karlsruhe in the autumn of the same year to study electrical engineering. I still hadn't really formed any precise notion of the work I would do afterwards and during the course of my life.

2 Karlsruhe – the Ray-Transformer

Karlsruhe's Polytechnic, known as the 'Fridriciana', is probably the oldest in Germany and has a very good reputation. Heinrich Hertz was one of many who had worked and taught there. I estimate that, in my time, there were about three to four thousand students in Karlsruhe. We cannot therefore regard the German Universities of that time as the student factories we know today, where student numbers of 20 to 30 thousand or more are the norm.

The relations between students and tutors were excellent and of a very cooperative nature during my time in Karlsruhe. I especially remember Professor Schleiermacher who taught us theoretical electrical engineering. He was a friendly old man. We also had a very fine mathematics professor called Böhm.

Professor Wolfgang Gaede taught us physics; he was one of the high gods and a little more distanced from us students. However, as mentioned previously, it was all very harmonious and we had no problems.

I found the teaching first-rate and well balanced. Professor Richter's lectures on the theory of electric machines were much influenced by the practical facts of engineering. We learnt a great deal about direct current machines, commutators and similar things which have now almost completely disappeared. We also had exemplary teaching in mathematics, chemistry and physics. Overall, it was pretty well balanced and had an academic flavour. It contained much more than just the purely practical aspects of engineering.

Spannhake, a teacher of worth, taught us about hydroelectric power machines. He was of a more practical bent. Professor Tolle taught us technical mechanics and he was very good, and Professor Nusselt was our thermodynamics lecturer.

The most important part were the lectures. We didn't have special seminars for our free subject, instead we would have a lecture on, for instance, Einstein's theory of relativity. The laboratories too were excellent. For our laboratory work we would be divided up into groups and given practical problems which we had to solve under the supervision of assistants. We worked quite independently. Later on we also had to design and build electrical machines. Our education was versatile and of good quality.

However, it was a shame that I no longer had the opportunity to study more physics. During my time in Karlsruhe, collaboration and communication with the physicists was not as good as it is today. There were few conferences, symposiums or meetings, and I also had very little personal contact with the physicists. Lectures on physics (Gaede) were of course included in our course, but we did no practical work.

It was also in Karlsruhe that I wrote my first publication – on a subject which has nothing to do with engineering. Inflation was rampant when I went to Germany in 1920; the value of the German Mark was constantly dropping. Price increases caused everyone to be interested in economics, and I would therefore make a daily plot of the US-dollar rate. This was for purely practical reasons. My father had initially bought me German Marks and now I wanted to know the best time to change money again.

This resulted in a dollar curve which, drawn on logarithmic graph paper, reached from the floor to the ceiling of my room. At first the dollar equivalent rose at a more or less linear rate, although naturally with major fluctuations, but by the end, in 1923, the exchange rate increased in such an alarming way that I had to use double-log graph paper. While one US-dollar had been the equivalent of 192 Marks in January 1922, by the end of 1923 it was about 4,200,000,000,000 Marks! This curve prompted me to write an essay for the Norwegian State Economics Magazine which was published in 1924 [Wi24]. I didn't take much notice of such things later on, but it was my very first publication.

Karlsruhe had a Nordic Club. Quite a few Norwegians and Swedes as well as a few Finnish students (Swedish and Finnish Fins), frequented this Club. There was also someone from Iceland

18

and a Dane, Mr. Hansen. We often held parties as there were many National holidays to be celebrated and there was much Cognac and Swedish punch to be had.

Some of the names have stayed in my memory; a Norwegian called Rotheim. He was the inventor of the spray-box, but his sole reason for inventing it had been to spray wax on skis. When he returned to Norway some time later, he had a batch of these spray-boxes manufactured. He had them patented as well, but it was not an economic success. He died quite young.

I also remember Jack Nilsen, a Norwegian tennis champion. He later became head brewer at Ringness. I bought his bicycle when he went back. Grude von Stavanger was a great baritone. There was a student of architecture called Björnson-Langen. His mother, the daughter of the Norwegian poet Björnstjern Björnson, had once been married to the publisher Langen (Simplicissimus) in Munich. He was great fun. And there was also my good friend Kaare Backer, he became a construction engineer, is still alive and over 92 years old. I went to visit him in February 1991 on the occasion of his diamond wedding anniversary.

I did a month's work experience in Strasbourg, in an electric motor factory. I had to wind the coils of a motor, a difficult task, and then I had to go to work outside, to connect various electrical cables onto a mast.

My diploma-dissertation, completed in 1924, was concerned with 'Potential Distributions in Chain Isolators' for high tension lines. This involved various problems. We had a tutor in high voltage technology, Professor Bonte, who had written a book which included several of his calculations for electric potentials. I had discovered that one of the calculations was wrong. This was the starting point for my dissertation, and I corrected his mistakes. I remember that I used differential calculus, but I also wanted to investigate the matter experimentally. I built a model of an overhead pylon at scale 1:100 with some suspended isolators and put it in a bath tub which I used as an electrolytic tray. As far as I can remember, this method was already known at that time, and

19

this is how I was able to measure the voltage distribution in water. After I had solved a few problems of surface resistance (silver electrodes), the thing worked quite well. This work was awarded with a 5.9 (6 was the top mark).

I had a lot of help when I was working on my dissertation in Karlsruhe. Existing equipment was made available to me, and I was allowed to use the workshop. Whenever it appeared necessary, my deadlines were extended.

In the autumn of 1922, while in Karlsruhe, I had already developed the basic ideas for a 'ray-transformer'. This machine would accelerate particles as if very high electrical voltages were available, but without the need for such dangerously high voltages, which could not be achieved in practice anyway.

The question I asked myself at that time was whether electrons in a ring shaped vacuum chamber would behave in the same way as if they were in a copper wire of an ordinary transformer's secondary coil. When the electric current in the primary coil changes, they should really be accelerated in the same way as the electrons in the transformer's secondary coil.

For example, if the alternating current in a transformer's primary coil changes direction 50 (or 60) times a second, this produces a force on the electrons in the secondary coil which 'accelerates' them in either direction. A single acceleration in one direction therefore happens within a fraction of a second and this was exactly the effect I wanted to exploit.

As the electrons were no longer confined within a copper wire, I had to switch on an appropriate magnetic field to keep them on a circular orbit. This magnetic field would, nevertheless, have to adapt itself to the increasing velocity of the circulating particles.

If there is a sufficiently high vacuum in the tube (imagined as the transformer's secondary coil), there should be hardly any electrical resistance and the electrons would achieve an extremely high speed within a very short time. This would correspond to the acceleration produced by a very high voltage. It was not so easy however to calculate the speed reached by these electrons. I was

20

soon convinced that the electrons would not take long to come close to the speed of light and that the formulas of classical mechanics would therefore no longer apply.

In those days, people were still not quite sure whether the formulas contained in Abraham's absolute-theory were correct or those of Einstein's theory of relativity. Because of this, I initially calculated the movement of the electrons in the ray-transformer on the basis of both theories. Later on I used only Einstein's formulas, as, in the end, these did appear to me to be better.

I came to the conclusion that acceleration within one rise of the current, that is, within less than a hundredth of a second, would be equivalent to a 'potential kick' of several million volts. The relatively small kicks at each revolution just kept adding up, and eventually resulted in this high number. It really was an amazing result, as this meant that the size of a machine which could reasonably be built would be quite modest; the electron orbits would be approximately 10 to 20 cm in diameter, if one were to use the technology for building transformer magnets which was available at the time.

In my first sketch (Fig. 2.1) I simply placed a flat (evacuated) accelerator vessel between the poles of a magnet [Wi23]. For this device I calculated the attainable energy. In a slightly later drawing (Fig. 2.2) I took into consideration that a second, independent, magnetic field is required to guide the electrons on reasonably steady orbits. This second magnetic field is induced by a second coil which can clearly be seen on the drawing.

After thinking it over for some time I arrived at the conclusion that there is an important relation between the accelerating field (of the transformer) and the deflecting or steering field (for the circular orbits), which must be maintained over the entire accelerating process – if one wishes to sustain the same size of the orbit during the whole acceleration process: The mean field within the circular orbit (that is, the 'accelerating' field) should always stand in a very particular ratio (precisely 2:1) to the deflecting field. This relationship, which later came to be known as the 'Wideröe

21

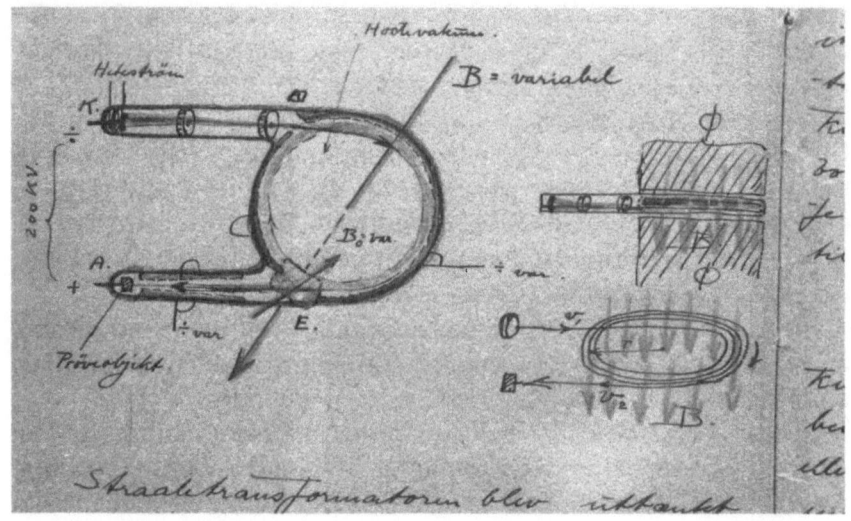

Fig. 2.1: The first sketch in Rolf Wideröe's notebooks [Wi23] of the ray-transformer.

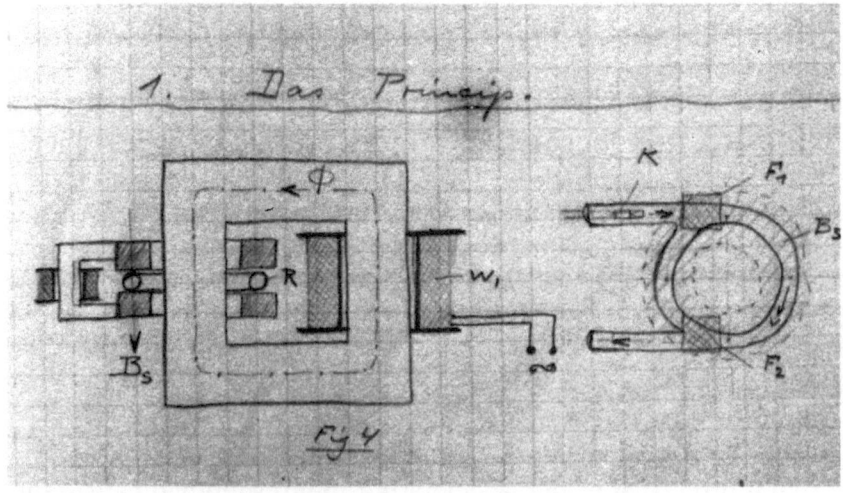

Fig.. 2.2: A further sketch by Rolf Wideröe which explains more precisely the operation of the ray-transformer.

22

relation', even permits both fields to be produced by the same primary coil, which again simplifies the whole machine. The magnet's yoke would be similar to that of an ordinary, largish transformer-yoke and could therefore, if the pole pieces had the right shape, provide both the accelerating and the steering field simultaneously. However, I had not quite got this far with my first ideas in Karlsruhe.

In the end, I had spent so much time thinking about the principle that I was convinced that it had to be correct – and this is really the crucial point: It is possible to accelerate particles with changing electromagnetic fields without using any static high voltages.

Until then the energy of charged particles had always been 'accumulated' by means of (static) electrical fields. Therefore, more and more 'volts' were required to achieve greater energy. What happens in a ray-transformer is, however, quite different and was quite new. Here the energy is accumulated in the form of kinetic energy, it can be increased without requiring high voltage. And this, in my view, was the important and basic idea for all further developments in this field and also for the entire particle accelerator technology which came later.

I didn't speak with anyone about my ideas and calculations in those days, because I realised that the 5th semester was too early to continue any work on this subject. I made a few notes on this in March 1923, which are still conserved in my copy-books [Wi23]. About half of my texts are in Norwegian, the rest are in German. However, after writing down these notes I put them on ice and continued my studies. I intended to proceed with this matter only at a later date.

At that time I knew nothing of what was going on in other laboratories, such as in England or Germany, where research on nuclear physics was being done, but I must have continued to ponder Rutherford's nuclear reactions and the possibility of creating better experimental conditions for them. In any case I wrote in one of my notebooks that one "would require at least 10 million volts and considerably more" to smash heavier atomic nuclei.

23

Rutherford's alpha particles attained a maximum of about 10 MeV. Furthermore, one would have to be able to shoot, under controlled conditions, a far greater number of particles onto the atomic nuclei to be smashed.

In my opinion, building a device similar to a ray-transformer was the only way to accelerate particles to much higher energies or, using the language of that time, to achieve the appropriate 'high tensions or potentials'. The energy of the particles was already then referred to as the 'potential' which would be required to accelerate them to that extent, independently of the actual method used to accelerate them. This is exactly where the energy units which are still used today, the electronvolt (eV), the kiloelectronvolt (keV) and the respectively higher ones (MeV, GeV and TeV) stem from.

Then I took a description of the ray-transformer to a patent office in Karlsruhe and requested that they apply for a patent based on my notes. However, I heard no more from them and when the work for the ray-transformer in Aachen started to go wrong I wrote the whole thing off. Many years later, it must have been in 1943, during the War, my travels took me back to Karlsruhe and when I searched for the patent office I discovered that the entire neighbourhood in which it had been located no longer existed.

In 1924, after finishing my dissertation and examinations in Karlsruhe, I returned to Norway where I first completed my practical work, which consisted of six months working in the locomotive workshop of the Norwegian State Railways. I also did my National Service in 1925 during which I commanded six men and a farmer with a horse and cart for 72 days! It was a wonderful summer. I came back to Karlsruhe during the autumn of 1925. First of all, I compiled all my ideas and calculations on the ray-transformer and took them to Professor Schleiermacher who, as already mentioned, taught theoretical electrical engineering. He was very nice to me and carefully read my manuscript. Then he said to me, "Your entire thesis is right here". He had given the whole thing a very positive verdict.

Then I went to see Professor Gaede who was responsible for physics, and showed my manuscript to him as well. When I returned to him a few days later I was rudely awakened. He believed that my proposed apparatus would never work and told me that I should forget all about it. Even with the best vacuum achievable at that time (I guess it was about 10^{-6} millibar), so many gas-molecules would be left over that the electrons on their long journey (covering several million kilometres in the small chamber!) would be absorbed far too quickly – far quicker even than it was possible to accelerate them. Of course, this was very sad for me and I was extremely disappointed.

However, I knew where I could find out more about the problem of electron absorption in gases. Professor Phillip Lenard, who had already been awarded the Nobel Prize in 1905, was working in Heidelberg at the time. His investigations were described in a book entitled 'Quantitatives über Kathodenstrahlen aller Geschwindigkeiten' [Le18] which I found in the library. Lenard had measured the scattering and absorption of electrons of several energies (from ten to one million electronvolts) in layers of matter, especially in air. I drew the results of his measurements on logarithmic graph paper and found a beautiful curve for the absorption as a function of the electron energy.

Accordingly, Gaede's assumptions were wrong. The losses due to absorption quickly decrease at higher electron energies (somewhere above 400 electronvolts) and after that they hardly matter any more. Yet this does result in a lower limit for the beginning of the acceleration in the ring; that is, a certain minimum of energy is required to inject the particles.

I did not, however, go back to Gaede. I had come to the conclusion that my original idea of writing a thesis in Karlsruhe was no longer feasible. My aim had been to build a ray-transformer, or at least an accelerating tube. Gaede would not have permitted me to do this. After thinking about it for a little longer, it also seemed to me that the technology available in Karlsruhe was not sufficient for my plans.

25

I used to like reading 'Archiv für Elektrotechnik'. In this magazine Professor W. Rogowski and Dr. Flegler had published papers describing their research work, for which they used very fast cathode ray oscilloscopes they had developed in Aachen. In their laboratory high frequency and high vacuum technologies were nurtured and therefore it was the right place for me. I wrote a letter to Professor Rogowski and asked whether I could work with him in Aachen and I received a warm reply. He wrote saying that he would be going to Switzerland for a holiday on such and such a date, and that he would be passing through Karlsruhe on the way back, "Join me on the train. We can travel together to Mannheim and you can explain it all".

I followed his instructions and we travelled to Mannheim together. The journey took about one hour. I don't believe that he understood much of my explanations, but I mentioned several times that I wanted to build a 'transformer' for six million volts, and that must have hooked him. He was ambitious and always wanted to be just that little bit in advance of the competition. Thus he said, "This sounds very good, come to Aachen and we'll sort it out".

So I moved to Aachen. On the eve of my departure we had a tremendous party. It ended with us hanging all the chairs on the wall. In the middle of the night, or rather in the morning, I rode off on the train. My landlady was appalled when she saw the state of my room, but my friends ironed it all out again.

I was well received in Aachen. I registered with the Polytechnic, was able to sit in on a few lectures and worked in Rogowski's laboratory.

3 Aachen – the First Operational Linac

Aachen was a rather unconventional place to work in. There were several assistants and PhD students who were investigating travelling waves, their penetration into transformer coils and suchlike. Dr. Flegler (he later became a professor in Beijing) was the head assistant.

In Aachen I met Ernst Sommerfeld. He was developing a small cathode-ray-oscilloscope under Rogowski's direction. Ernst was the son of the famous physicist Arnold Sommerfeld (see for instance [Ec93]). We became great friends and have frequently had the opportunity to get together again since and throughout our lives. He later specialised in the field of patenting, and before the War lived in Berlin where he worked as a patent agent for Telefunken. During the War he was called up and became an officer's driver for a while. He moved to Munich after the War, where he lived in his father's house and started his own company. Most of my patent applications (there were over 200 in all) were looked after and submitted by him.

Ernst often came to visit in Norway and we made several tours to the high mountains. During my period in Hamburg between the end of 1943 and March 1945 I visited him a few times and he also came to see me later on in Baden. Sadly, he died of a stroke in 1980. His father Arnold had been teaching in Aachen and had worked there for several years and I suspect that this was the reason why Ernst was working with Rogowski. Arnold Sommerfeld later went to the USA, and therefore I was able to get early information about Lawrence's work as well as the development of the cyclotron. However, I did not meet Ernst's father until many years later in Zurich where they had come to visit us.

In Aachen we had the opportunity to hear some very good lectures on electrical engineering by Rogowski, and on aerodynamics by Karman, who was later to go to California. We used to play tennis with Karman's assistants. The biggest departments in the Polytechnic were the metallurgy departments, this was primarily due to the Rhineland's industry and mines. Incidentally, I was the only Norwegian in Aachen during my time there.

I was soon busy building the ray-transformer. I believe that my workshop activities at the time were paid for by an institution for German Science called 'Notgemeinschaft der deutschen Wissenschaft'. Fig. 3.1 shows my working place in the institute's cellar. The dimensions demonstrate how little space there was.

The city's power station supplied me with an iron yoke. It had been taken from a relatively small three-phase transformer and was about one metre tall. I had part of the yoke cut off in order to obtain a simple iron return path, that is, a two-phase transformer,

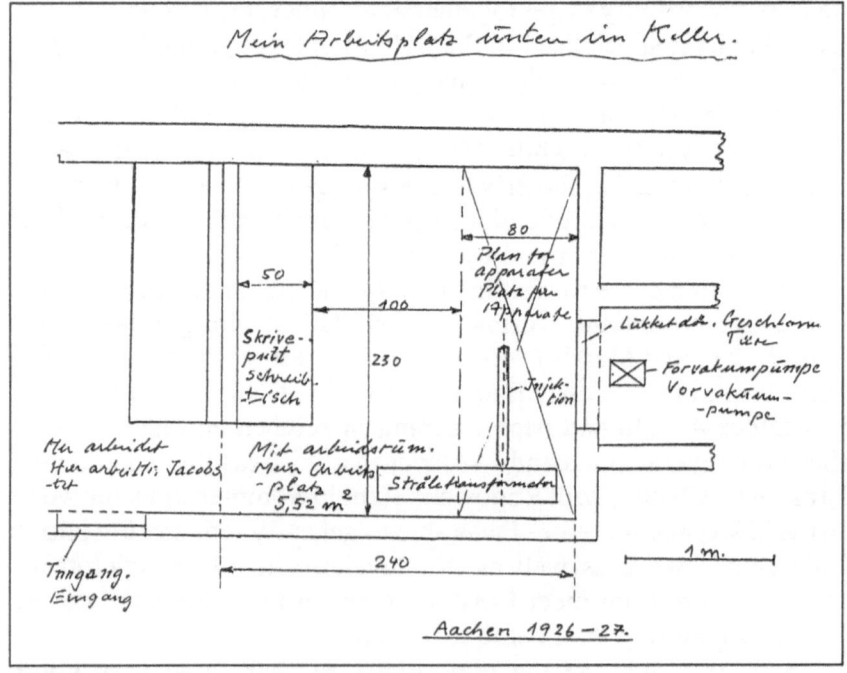

Fig. 3.1: Wideröe's working area in Aachen.

and then I took out a piece to obtain two poles at the top. I used small iron plates to shape the induction and steering regions between the two pole areas. The drawings are shown in Fig. 3.2 and 3.3, and are excerpts of my dissertation.

The poles were shaped in such a way that the magnetic fields in the accelerating and deflecting regions followed the 2:1 ratio which I had already discovered in Karlsruhe, and which today is named after me. Of course, I had also made use of the simplification which is a result of this ratio: both the accelerating and deflecting fields were induced by the same coil. The correct ratio is provided by the shape of the magnet's poles. I had measured the fields between the poles quite accurately with test-coils and verified that they complied with the 2:1 ratio.

We had an excellent glass-blower in Aachen, for it was not the easiest of tasks to make the vacuum tight ring tube. The glass-ring was about 15 cm in diameter, and the tube had a cross section of 15 mm. It was fitted with a ground glass connection for the injection tube. The ring stood upright and the electrons were injected from above, as can be seen in Fig. 3.2. A vacuum pump was connected through another glass tube.

To produce and inject the electrons I used a source which was similar to those used in the cathode-ray-oscilloscopes of Rogowski and Flegler. It was quite a reasonable source of electrons; the electron beam was then focused by a long coil and there was a small entrance-slit, which I could open and close from outside.

During the early phase of my experiments, I shot the electrons into the evacuated glass ring tube with a weak starting field. Then I turned on the magnetic field by switching on the current and at the same time attempted to observe the accelerated electrons. The internal walls of the glass tube were covered with a fluorescent material which was supposed to give some fluorescent light when it was hit by electrons. In this way I hoped to observe some of the electrons after they reached their highest energy.

In theory, the electrons were supposed to reach an energy of up to 6.8 MeV, which, with a normal voltage generator, would have taken 6.8 million volts to achieve. At that point I had to lead the electrons away from their nominal path, that is, I had to 'extract' them from their orbit, if I may put it this way. The coils of the magnet had a fuse. When the current reached its maximum, the fuse turned off the current and simultaneously turned on the current in another coil which was supposed to kick the electrons against the walls of the ring tube. It was all rather primitive and I described everything very precisely in my notebooks.

I fired the magnetic field many times by shutting the switch which is also shown in Fig. 3.3, but I could not see any accelerated electrons (there was no fluorescence on the inside of the wall). Of course, fluorescence is a rather poor method for detecting electrons, and I am sure that a good physicist would have thought of a much better way to do this.

Later on it became clear that it is possible to make both the test set-up and the measurements much simpler by exciting the magnetic field with alternating current, which was how I had planned it in my original sketches (instead of having to resort to awkward switching on and off). Well, I never got that far.

I had made no provisions for avoiding the effects of electrons which deposited on the internal walls of the ring. As I was soon to find out, 'islands' of electrons formed in some places on the internal walls of the ring. They had an important role to play. These islands formed wherever the wall was hit by electrons running out of their nominal path. They produced an electric potential which reduced the energy of the injected electrons by about one third. I therefore had to adapt the field to this lower energy during injection. I had a faint hope that the charges on the walls would produce some stabilising forces, but this was not the case. However, I did finally manage to get the electrons to circulate in the ring approximately one and a half times.

Later on, the charge-islands were avoided by coating the inside wall of the ring with a slightly conductive graphite layer. If I compare all this to my experiences in Hamburg between 1943 and 1944 and in Baden after 1946 at BBC, I can say that it was not only

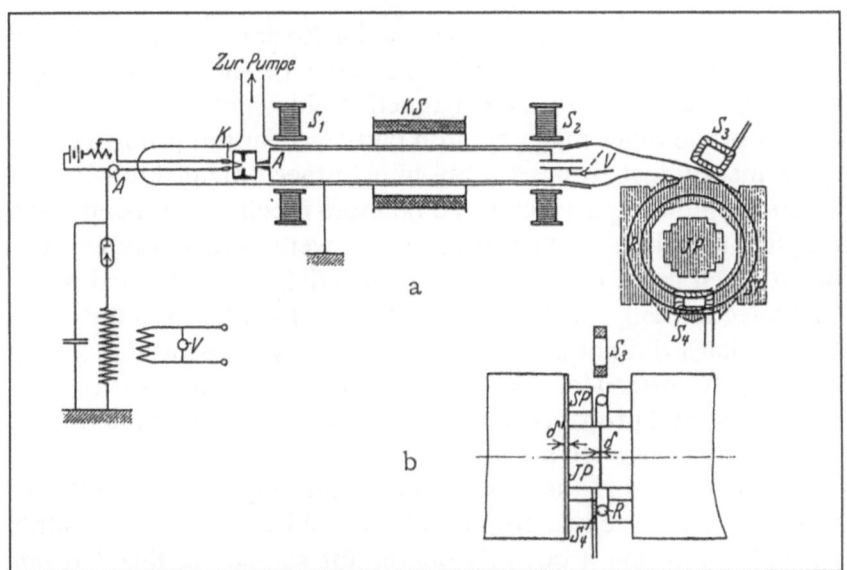

Fig. 3.2: Diagram of the Aachen ray-transformer [Wi28].

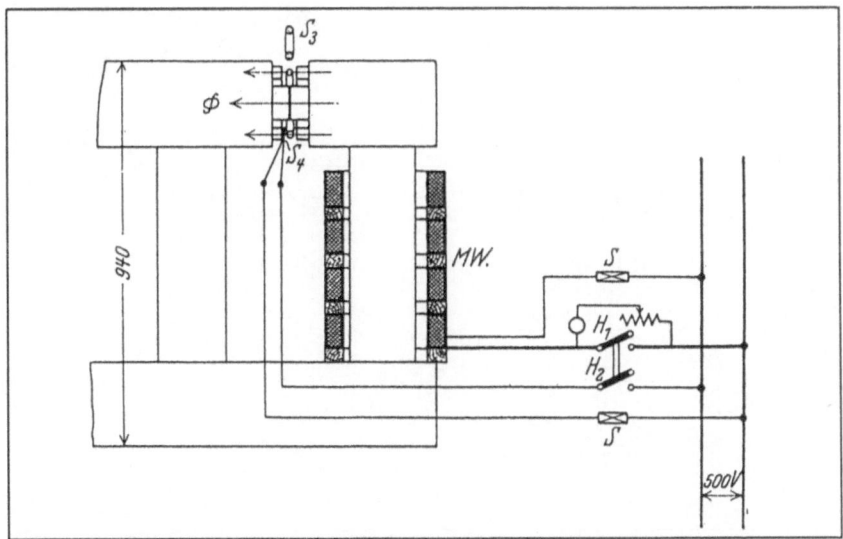

Fig. 3.3: The experiment set-up for the ray-transformer [Wi28].

the omission of a conductive wall coating (to draw off the electrons from the walls) which denied the machine it's success. The shape of the iron core (and thus the magnetic field which was created by it) and of the other magnetic iron parts was far too primitive and quite insufficient to meet a ray-transformer's (later known as a betatron) high requirements. To be more precise: The conditions required to stabilise the electrons' orbits were as yet unknown, and my Aachen machine was far short of satisfying such conditions. The injection too, was less than sufficient. I think it was fortunate for me that I did not continue with those ray-transformer experiments, but instead stopped immediately. My own insufficient experience and probably the conditions in Rogowski's laboratory were simply not adequate to the task.

When I realised that I was not having any success with the machine, I reported to Rogowski. He told me that he couldn't possibly grant me a doctor's degree for something that did not

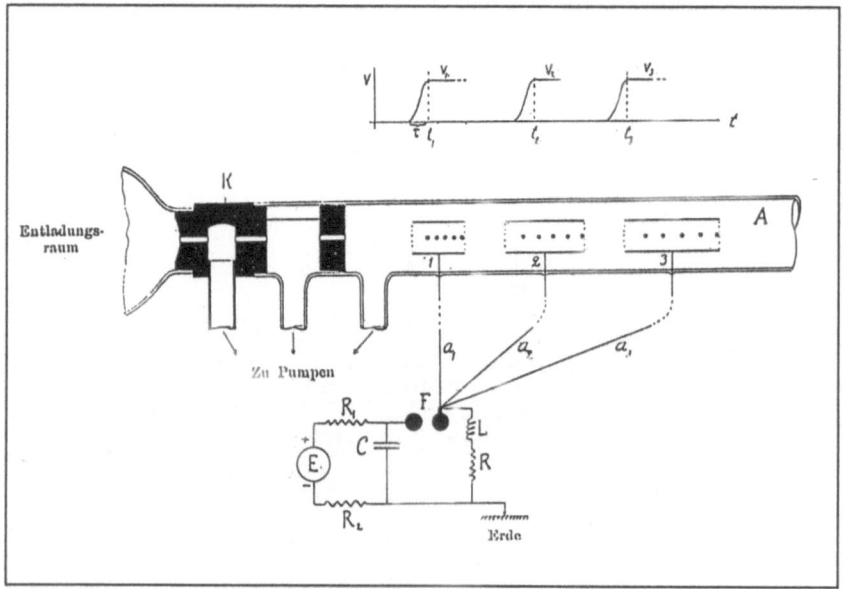

Fig. 3.4: Ising's first suggestion for a linac [Is24].

function. I was well aware of this, so I had to construct something that would work – and I already had a solution in mind.

As part of my reading in the Karlsruhe library I had come across a publication by Professor Gustav Ising in the Swedish magazine 'Archiv för Mathematik, Astronomie och Fysik' [Is24]. In this article he proposed that electrons should be guided through a straight vacuum tube, inside a series of metal tubes ('electrodes') in which a so-called travelling wave was produced by high frequency alternating voltages. These voltages would be applied to the tubes through adequate delay lines. Fig. 3.4 shows Ising's original drawing. The particles would be accelerated as if they rode 'on the front of the wave', in Ising's tube. I committed this article to memory and thought at the time that I may be able to make something useful of it one day, especially if my ring ray-trans-former didn't work.

However, I already understood something about travelling waves and the many possible problems associated with them. The electrodes suggested by Ising, as sketched in his publication, would have reflected these waves, and I could see that it would not be possible therefore to produce any accelerating voltage. However, the basic idea was very interesting, and I developed from it the so-called 'drift-tube'. This simple tube was connected to a high frequency voltage supply and (having the appropriate frequency and length) would accelerate electrically charged particles two times, namely once as the particle entered the tube, and a second time as it exited (see Fig. 3.5). While the particle is inside the tube, the voltage is reversed without affecting its motion.

Electrons are not particularly suitable for this type of accelerator. They rapidly reach such high speed that one would require either a very long tube or a very high frequency for the alternating voltage. At that time (1927), it was not possible to produce sufficiently high frequencies for such apparatus; at most one could perhaps count on a few megacycles, which is not enough.

Because of this I resolved to try the 'drift-tube' principle with particles which were heavier and which would move at a much

33

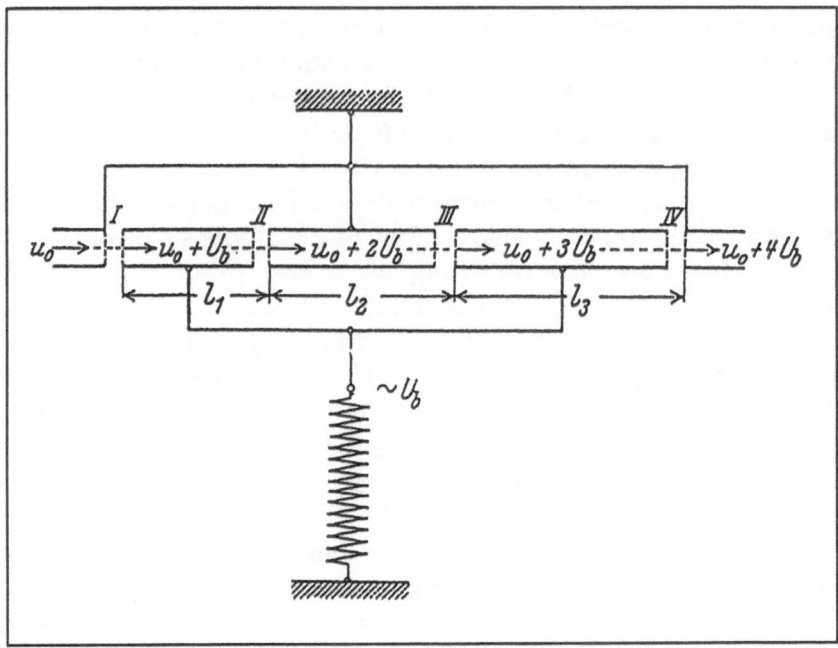

Fig. 3.5: The principle of the 'drift-tube' as illustrated in Wideröe's thesis [Wi28].

slower speed. I decided to use potassium and sodium ions, that is, potassium and sodium atoms which, because a few of their electrons are missing, have a positive charge. I am referring therefore, to so-called 'anode-rays' which had already been known in physics for quite some time.

One of my tennis partners worked at the Institute of Metallurgy and he came to my aid, building the activator for the anode of the Kunsman-type which I used in order to produce the ion beam for my little accelerator. After that, the rest of the equipment was quite easy to construct. It was housed in an 88 cm long glass tube. A diagram of the installation taken from my thesis, is shown in Fig. 3.6. If I remember rightly the accelerator cost no more than four to five hundred Marks.

The ions went into the drift-tube at relatively low speeds. As they entered, they received a first voltage kick of up to 25,000 volts and as they exited a second one of approximately the same value. The voltage was reversed at just the right moment, when the ions were inside the tube. After this, the ions passed through a second tube which was not connected to the high frequency voltage, it was earthed. Then they moved between two electrically charged plates where they were deflected more or less, depending on their speed. Finally they reached a sensitive photographic plate of a type which in those days was already in use to make X-ray photographs. The accelerated particles 'exposed' the emulsion's silver bromide grains (just as light would) and formed narrow stripes which I could measure after I developed the plates.

Following a few calibrating measurements, the ions' final energy for each accelerating voltage was precisely determined. The readings taken with the potassium and sodium ions showed

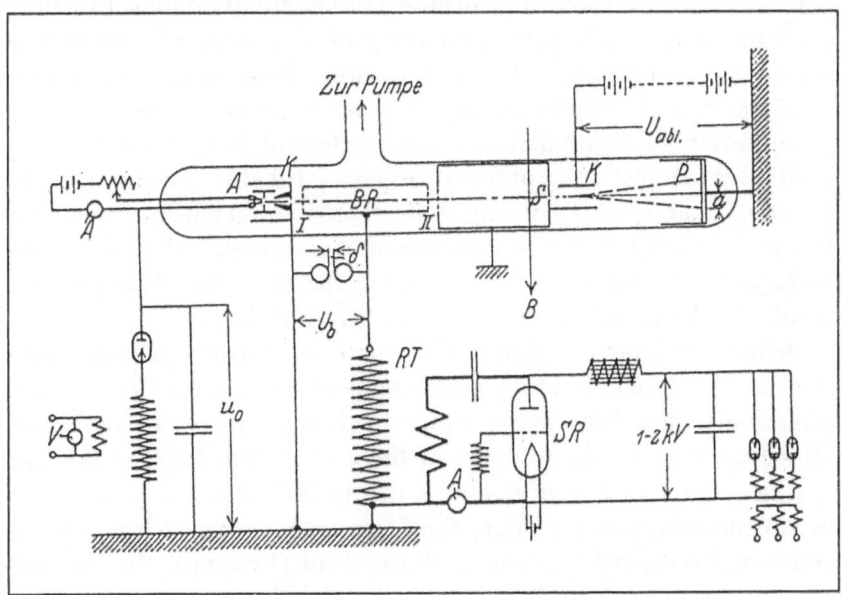

Fig. 3.6: Acceleration tube and switching circuits [Wi28].

that everything was functioning as planned; the ions really were accelerated twice by the same high frequency alternating voltage and finally achieved a speed for which one would otherwise have required 50,000 volts! For the first time it was thus proven that it is possible to accelerate electrically charged particles several times using high frequency alternating potentials. It was therefore possible to accelerate particles as if one had available very high voltages without, however, having to take recourse to a correspondingly high voltage device.

There was also no reason to doubt that my procedure could be repeated as often as desired using a sequence of such drift-tubes in order to accelerate the particles to even higher energies. In principle, it was possible to 'extend' them indefinitely to achieve ever higher energies. In fact there is today such a linear accelerator at Stanford University in California, which, over the years, has been extended until now it is approximately 5 km long. It accelerates particles as if 50 thousand million volts were available. My little machine was a primitive precursor of this type of accelerator which today is called 'linac' for short. However, I must now emphasize one important detail. The drift-tube was the first accelerating system which had earth potential on both sides, i.e. at both the particles' entry and exit, and was still able to accelerate the particles exactly as if a strong static electric field was present. This fact is not trivial. In all naivete one may well expect that, when the voltage on the drift-tube is reversed, the particles flying within would be decelerated – which is clearly not the case.

After I had proven that such structures, earthed at both ends, and in which acceleration could take place several times, were effectively possible, many other such systems were invented. However, I will refer to some of these at greater length later on.

There are exact reproductions of my little Aachen installation in various museums, namely the German Museum (Munich), the German Röntgen-Museum in Remscheid (Lennep), the Norwegian Radiumspital in Oslo, the Norwegian Technical Museum in Oslo, the Swiss Technorama in Winterthur and the Smithsonian

Institution, Washington DC, USA. It must be said, however, that the reproductions are more beautiful than the original I built in Aachen. These models (which, with the addition of a few components are even capable of functioning) were built in 1982 in the Radiumspital in Oslo. Their construction was suggested by a friend who worked there, the physicist Olav Netteland. Regrettably, before work could begin he suffered a serious stroke. We therefore tried at first to have the models built at BBC in Baden, but this proved to be too expensive. In the end they were made by an apprentice at the Radiumspital in Oslo, exactly to my specifications. Another similar model is now being built at the research centre DESY in Hamburg, also in the apprentices workshop.

The important invention however, was the drift-tube, driven by high frequency voltage. It supplied the foundation for the development of particle physics with high energy accelerators, particularly with reference to the ideas which arose for the cyclotron and for the synchrotron. The principle of the 'synchrotron', using a bent drift-tube, for example, was patented by myself in Norway in January 31, 1946 [Wi46]; a facsimile of this patent is reproduced in Appendix 2. Moreover, my original simple drift-tube was the starting point for the development of all later variations of 'accelerating cavities' used in circular as well as linear machines. Of course I made a big mistake when I did not have the drift-tube immediately patented in Aachen.

Rogowski took hardly any notice of my work. I don't think that he ever as much as looked at my linac. It was expected that my thesis would be published in a periodical and I had no problem getting it into 'Archiv für Elektrotechnik' [Wi28]. The publication is almost identical to my thesis; only the Lenard curves are missing. Rogowski and Professor L. Finzi (physics) were my examiners. I had no problems there either and I finally obtained my title of 'Doktor-Ingenieur' on November 28, 1927.

It is not that easy to write such a doctoral thesis. I was given no instructions and wrote everything myself. In my thesis I also mentioned a few methods and principles for achieving higher voltages with potential-fields, for example Marx generators (a set of parallel and series capacitors) and similar installations. Unfortunately, there were a few printing mistakes in the thesis, but these were corrected in the English translation which was not written until about 1965. This was when I was a consultant at DESY and, as I clearly remember, many people helped me with the translation, including G. E. Fischer, F. W. Brasse, H. Kumpfert and H. Hartmann. This translation appeared in the book 'The Development of High-Energy Accelerators', which reprinted important publications on this subject [Li66]. I did have a few problems with Stan Livingston who was editing the book. He wanted to publish only the section on the functioning linac, so I had to battle with him and said, "either you take the whole thing or nothing at all". In the end he accepted it in its entirety, including the piece on the ray-transformer.

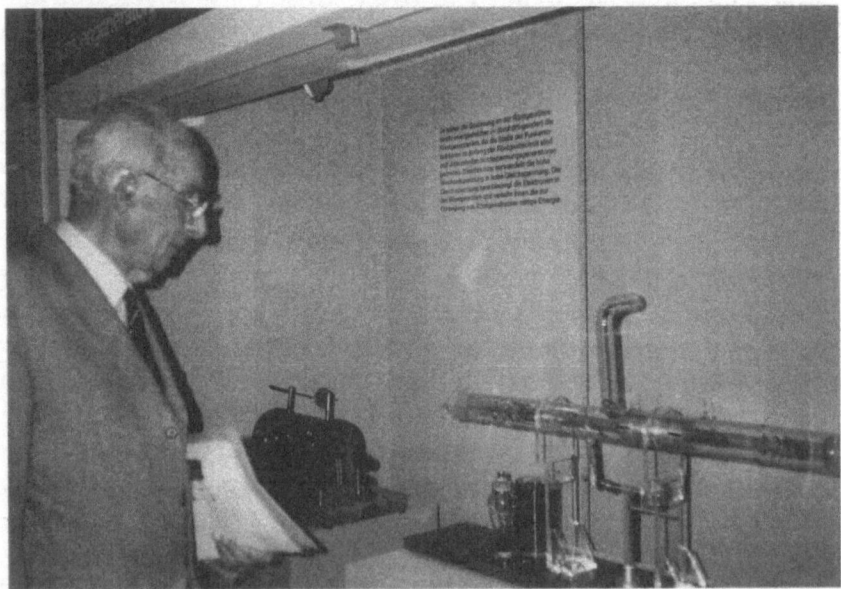

Fig. 3.7: Rolf Wideröe in front of one of the linac models in the Röntgen-Museum in Remscheid, photograph by Ragnhild Wideröe.

4   Cyclotrons and Other Developments

At this stage I would like to say a few words about Ernest Lawrence's work in America. Lawrence was of Norwegian extraction and his family name had originally been Larsen. He was a very interesting person, spirited, stubborn and full of enthusiasm. Furthermore, he had a definite thirst for adventure.

Lawrence once recalled in my presence that he had been at a conference in Berkeley (it must have been in 1928) where the presentations became rather tedious for a while. He therefore removed himself to the library and found my thesis in the magazine 'Archiv für Elektrotechnik'. He looked at the pictures and formulae only, as he could understand little or no German. From these illustrations he gained an immediate understanding of my drift-tube principle. However, it was of great advantage to him that he didn't know the German language; he could not understand my reservations on the stability of the orbits in circular accelerators, as included in the essay.

Thereupon, Lawrence, who worked in the then 'Radiation Laboratory' in Berkeley near San Francisco in the USA, together with his student David Sloan, built first a linear accelerator for Mercury ions with a total of fifteen tubes, and later one with even more [La31a], in exact accordance with the principle sketched in Illustration 3.5. He was thus able to accelerate ions to an energy of 1.3 MeV, i.e. as if he had 1.3 million volts at his disposal, although he in fact used only 48,000 volts of high frequency voltage. It was a tremendous achievement!

However, Lawrence was already suggesting that the drift-tube should be transformed into a D-shaped box and that the particle paths should, with the help of a magnetic field, be 'wound up' into a spiral. Thus he had invented the famous 'cyclotron'. He had discovered that, although the radii of the particle orbits increase as

Fig. 4.1: Diagram of the first cyclotron by Lawrence and Livingston [La31b] [Li62].

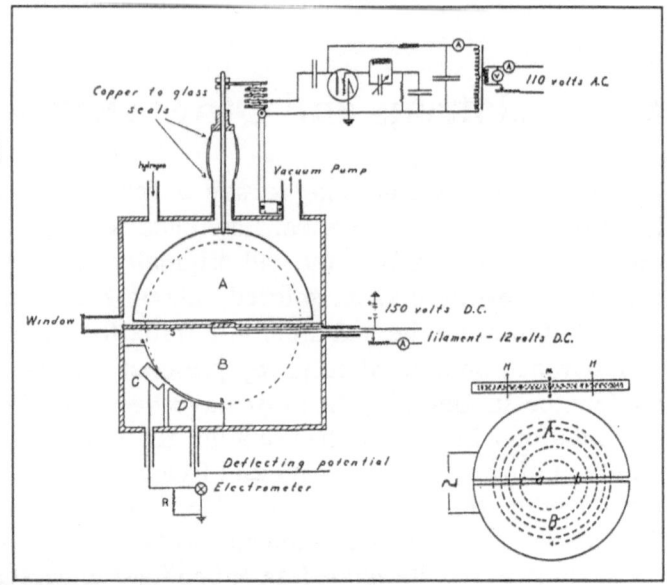

Fig. 4.2: Photograph of parts of Lawrence and Livingston's first cyclotron [La31b] [Li62].

the energy grows, they require the same amount of time for each revolution, because their speed also goes up. Therefore, the frequency of the accelerating voltage could remain constant (although only as long as classical mechanics remained sufficiently accurate) and this greatly simplified the installation. He published these ideas with his student N. E. Edlefsen [La30] even though the first experimental tests were not at all successful. He was very confident really!

However, I must now mention that Rogowski's assistant in Aachen, Dr. Flegler, had the same idea some time around 1926. During a meeting held to discuss work in progress, Flegler asked whether it would be possible to wind the ion paths into a spiral. I replied that it would be very difficult to stabilise the circular orbits, which is exactly what I later wrote in my thesis. That is how Flegler's suggestion for a cyclotron was abandoned and I was the one who more or less killed the idea (see also Box 6).

In contrast, Lawrence, together with Stan Livingston (another of his then students), pursued this same idea and, in 1930, constructed the first functioning cyclotron for protons [La31b]. All they had was a four inch magnet from the laboratory's stock, and, with this small installation, they could accelerate hydrogen ions to a modest 80 keV. However, this did definitely confirm the principle – and Livingston was awarded a PhD on its basis [Li31].

Their second cyclotron had a magnet with a diameter of 10 inches and with this they were able to accelerate protons to 1 MeV as well as perform experiments. Thus (with M. G. White) they confirmed the nuclear disintegration, which had previously been observed by Cockroft and Walton in England. The third cyclotron had a diameter of 27 inches and in 1934 it accelerated heavy hydrogen nuclei (heavy hydrogen had just been discovered in 1931) to 5 MeV, which corresponded to 5 million volts – here too without having to resort to such a high voltage!

Afterwards Lawrence went on to build several more, very successful cyclotrons, and in 1939 was awarded the Nobel Prize. It was the start of large accelerator development for nuclear and

Box 2

Cyclotrons and Synchrocyclotrons

Cyclotrons became the working tools of nuclear physics. Many were built throughout the world. They made it possible to smash atomic nuclei, just as Wideröe had dreamt in his youth; but they could also be used to produce useful quantities of new isotopes and for much fundamental research work. The energy of the accelerated protons (cyclotrons are not well suited for electrons) could easily reach 40 MeV, and it also became possible to accelerate heavier atomic nuclei. Of particular importance was the high number of particles (also called 'intensity') which could be accelerated with cyclotrons.

Subsequent attempts to achieve higher energies with cyclotrons were problematic because classical mechanical equations were no longer applicable; it became necessary to refer to the more precise formulae of Einstein's relativistic mechanics. However, this meant that Lawrence's original constant frequency idea no longer worked. As the particle paths' radii increased in size, the frequency had to be changed, it had to be adapted to the particles' relativistic speed.

Although this is possible in principle, it means that the frequency had to be changed during the acceleration process. It is therefore possible only to accelerate relatively small bunches of particles and the frequency has to be precisely adjusted in the process. The total number of particles thus accelerated is reduced by a factor of about one hundred. Yet this was accepted in order to achieve higher energies. These machines were called 'synchrocyclotrons'. Many of them were constructed later on and they reached energies of several hundred MeV.

The synchrocyclotron in Dubna (previously USSR) for example, which was first operated in 1954, achieved an energy of 680 MeV and was fitted with a gigantic magnet weighing 7,200 tons. However, there were also many smaller synchrocyclotrons with which important research work was undertaken.

With these machines it became possible to systematically investigate artificially produced 'mesons', whereby the field of nuclear physics was left behind and the next step forward, particle physics, was taken. The CERN synchrocyclotron ('SC') in Geneva became operational in 1958 and served several generations of particle and nuclear physicists.

particle physics at high energies. However, I didn't make this type of machine my particular business. This was partly because I was engaged on quite different activity at the time, but I did closely follow their emergence and progress.

I came to the conclusion that this was not the best route towards achieving higher energies. The spiral orbits within these accelerators require a magnetic field which covers a large area and is best produced with an iron yoke. Not a major problem, as long as the energies were not too high. If, on the other hand, one wanted to go to higher energies, a limit was very soon attained, which was given by the magnet itself, by its weight and its cost. My ray-transformers encountered the same problem. The magnet required to accelerate to higher energies would have been much too large.

Yet I hoped to keep the particles within a relatively narrow ring tube, as was the case in the ray-transformer, and still manage to accelerate them – possibly without the bulky inner part, the accelerating induction field. This would have had some advantages over the gigantic D's in the higher energy cyclotrons and my thoughts were therefore levelled in that direction. This remained a dream however; I did not seriously occupy myself with this subject until later, when, for purely personal reasons, I found time for it – and this wasn't until 1945.

Apart from Lawrence's cyclotron, the Thirties saw another important step forwards. This was thanks to the work of many physicists, but perhaps in particular to that of Louis Alvarez. He too worked in the Radiation Laboratory in Berkeley, which is today known as the 'Lawrence Berkeley Laboratory' (LBL). I should imagine that Alvarez developed his proposal on a line with Lawrence and Sloan's successful linear accelerator, because he became Lawrence's assistant in 1936.

With the advances of high frequency technology, Alvarez was able to build electrode systems in cylindrical boxes, in which resonant electromagnetic waves could then accelerate particles. Since then, two types of drift-tubes are distinguished, those of 'Wideröe' and those of 'Alvarez'. The latter have to be built into

43

Box 3

About Drift-Tubes and Waveguides

Specially shaped boxes, usually cylindrical, were soon developed, based on the tanks devised by Alvarez. Known as 'resonators', these could be made to oscillate at high frequency. Because both ends are held on earth potential, a standing electromagnetic wave is produced inside, which accelerates electrically charged particles as they fly past at the right moment.

Today, many accelerating cavities are built this way and are then used in both linear and circular machines (such as synchrotrons and storage rings). Acceleration corresponds to several hundred thousand volts per metre of resonator structure.

Nowadays, resonators, which have their internal surface cooled to 4 K are in use and, since they are made of particular materials like Niobium, they become superconducting. A higher accelerating voltage (gradient) is achieved (several million volts per metre during continuous operation) and heat losses are much reduced.

However, a second type of accelerating tube was developed in parallel, in which the high frequency power is introduced at one end and withdrawn again at the other. This causes a 'travelling wave' to form internally which is also capable of accelerating electrically charged particles flying through at the right time. This complies with Ising's original idea [Is24], but can only be made to work by providing the internal surface of the tube with a very particular shape – as Wideröe had already realized in 1927. A usual type of such tubes is called 'iris-loaded waveguide' (see Fig. 4.4).

It was possible to achieve acceleration gradients of 17 MeV per metre in normally conducting linacs of several kilometres length, like the Stanford Linear Accelerator. Almost twice as much is realized in small machines, such as those used today for medical purposes.

High frequency technology, which was developed to serve radar and television as well, is the prerequisite for operating all these accelerating devices. High performance transmitters with a frequency between 300 and several thousand MHz are utilized, thereby employing very large transmitter tubes with power outputs reaching the megawatt region. The electromagnetic waves are sent into the cavities through specially designed 'wave guides' (accurately shaped metallic tubes), instead of cables.

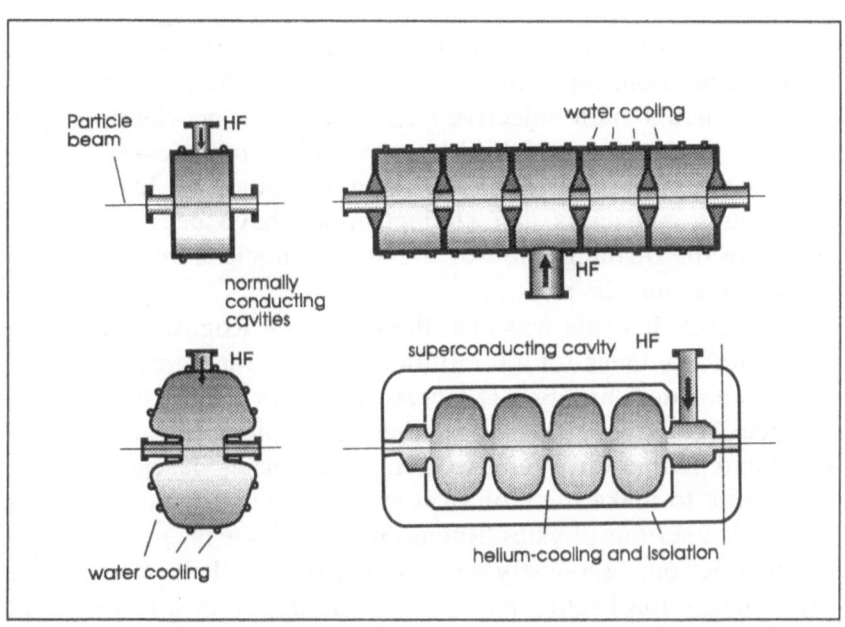

Fig. 4.3: Various types of resonators used in particle accelerators.

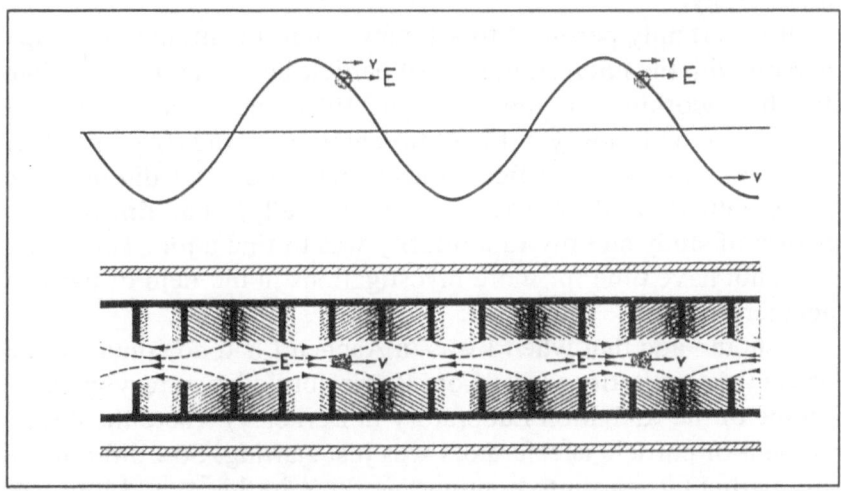

Fig. 4.4: A linear accelerator's 'iris-loaded wave guide' [Wi62].

an 'Alvarez-Tank' of very particular shape. In some modern linear accelerators both types of structure may even be applied.

Lawrence's main objective was to construct accelerators, particularly cyclotrons, and this he pursued like a man possessed. Yet the construction of larger and larger machines by his younger colleagues, assistants and students must have been motivated more by the disintegration of the atomic nucleus and other research into nuclear physics.

I suspect that this was also the case with Rogowski when he supported my ideas for a 6 million volts ray-transformer. He was a well educated, highly intellectual man with a most lively intelligence. We never spoke about these possible applications however, and neither did I refer to them in my thesis. It was probably premature to make mention of it at the time and would not have counted as serious physics. Rather, it would have been regarded as science fiction. I modestly wrote in my thesis, "It is possible that high energy ion beams may be of some importance to physics". Quite an understatement really, because ever since 1919, splitting the atom had been the leitmotif behind my interest in high voltage technology.

It is certainly pertinent to ask why I didn't continue to occupy myself with the interesting field of particle accelerators after I had finished working on my thesis in 1927 and 1928. Well, the cyclotron had not yet been invented and the first nuclear disintegrations with artificially accelerated particles did not take place until 1932. So it was quite simple really; I had finished my period of study and my first priority was to find a job. Therefore, I did not have time for more investigations in the field of particle accelerators.

I should add that when I was in Aachen I had no contact at all with other institutions (like Lord Rutherford's laboratory in Cambridge or the Radiation Laboratory in Berkeley) where the development of particle accelerators was just starting. So I did not see any particular reason to continue working in this area. Moreover I could not at that time think of any use for particle accelerators, other than splitting atoms - which I considered to be a far distant goal.

I wasn't particularly interested at that time in the option of using high-energy electrons to produce harder (i.e. more powerful or deeper penetrating) X-rays. Accordingly, I did not think of X-rays for use in either the investigation of materials or in medicine. I considered my work in Aachen as completed, and, for the time being, concentrated on other tasks.

5   Relays Are Interesting Too

Although Rogowski only pursued his own scientific work and did not accept commissions for research or development, he had good connections to industry. He recommended me to the director of AEG's transformer factory in Oberschöneweide near Berlin. His name was Dr. Stern and he later became a university lecturer. This factory needed someone to develop safety-relays, to protect power plants against short circuits in high-voltage transmission lines. Shorts like this could happen for many reasons, for instance, if a tree fell on a line. I went to Berlin in the spring of 1928.

I met Mr. J. Biermanns in the AEG transformer factory. A very nice man, he held an important position as the head electrician at the factory. Among his achievements was a book on 'Overshoot-Currents in High Voltage Installations' [Bi26] and he was later awarded a professorship too. I last visited him in Hanover during one of my drives through Germany at the beginning of the 1950s. He died shortly afterwards.

Together with Reinhold Rüdenberg, Biermanns had invented a relay to protect electric power plants. Rüdenberg was head electrician for the Siemens-Schuckert factory in Berlin-Siemensstadt, and as such, head of the 'Scientific Department'. He was considered an authority in the field of high current technology in Germany at the time. He wrote a book about relays and was interested in the problem of linked power stations [Ru29]. We never met. The physicist Max Steenbeck with whom he submitted a patent in 1933, to which subject I shall return later on, worked in his department.

The principle behind the Biermanns-Rüdenberg relays was that the short-circuit voltage was divided by the short-circuit current and that the delay-time of the oil-switch (interrupting the current in the line) could be set proportional to the impedance (i.e., the line's resistance). If several such relays react during a fault, the nearest relay in the high-voltage transmission system will react first and thus selectively switch off the fault. The relays also had to be directional, an important fact when dealing with parallel lines. However, Biermanns' relay was rather primitive. It had a relatively long reaction-time (delay) and very poor directional sensitivity. Biermanns' assistant Otto Mayr had proposed a different construction and it became my task to develop and construct the new relay.

Otto Mayr was from Kempten, the same age as I, and we became good friends. He later developed a pneumatic switch and a physical explanation for the switching theory. He was subsequently awarded an honorary Doctorate in Engineering. His last years were spent in Schwäbisch Hall where he died in 1989.

I worked first in the transformer factory and later in a relay factory called 'Dr. Paul Meyer', which had been bought by AEG.

My first years in Berlin were very interesting; it is such a stimulating city. In 1929, Berlin hosted an important international conference. I went along and heard lectures by Einstein as well as Eddington who reported on the stars' generation of energy by nuclear fusion. He spoke of temperatures of 40 million degrees.

My work at AEG was quite fascinating to me. I saw the relay as a kind of artificial intelligence as we would say today, or as a sophisticated analogue computer. During the time I developed relays I submitted a total of 41 German and 2 American patents for AEG. It was a very productive period.

At AEG I met Arno Brasch and Fritz Lange. On top of the roof of one of the factory's buildings, there was a high voltage Marx-generator with which it was possible to obtain over one million volts, and I suspect that Brasch and Lange used it to irradiate mice [Br30]. They also tried to induce some nuclear reactions, and may even have had some success, without, however, being able to provide exact proof. Brasch and Lange conducted a few other, rather hair-raising, experiments with high voltages. For instance, they wanted to 'divert' high voltage from storm clouds at Monte Generoso in Switzerland, which, of course, was very dangerous.

I also met Leo Szilard in Berlin. He was a very interesting man. I remember sitting in a cafe while he told me about one of his high voltage projects. He wanted to build several transformers, one on top of the other. The lower ones were to activate the ones above in some sort of cascade circuit. Szilard had many good, although often vague ideas. He was fun to be with, a typical Hungarian. Even then he had a good relationship with Einstein. I believe that together they developed the principle for a type of refrigerator and then submitted it for patenting – if, that is, my memory serves me right.

During my time at the AEG factory in Berlin-Schöneweide I was able to dedicate myself entirely to relay technology and was generally free of business and administrative tasks. I gave very little actual thought to particle accelerators at the time, but I did keep an eye on their development.

However one of my laboratory colleagues at the transformer factory, his name was Kujath, wanted to continue the development of the ray-transformer. We sat in the same room, he behind me, and I never saw him again afterwards. I remember explaining to him that if the magnetic stray-fields were correctly shaped, there have to be forces with which it may be possible to stabilise the electron orbits. However, according to my Aachen experiences and as I believed at the time, these forces would be inadequate for the task. This is more or less what I wrote in my thesis.

In those days I was beginning to think about a stronger way of focusing, that is, about improving the bundling of particles on their intended circular course. A practicable solution did not come to me, however, until much later. During my Berlin period I had more or less written-off the ray-transformer. This does seem rather strange to me now.

Then came the Depression in 1930 and the years that followed. I was running a laboratory and it was difficult and embarrassing to have to give notice to many of the engineers and employees. At first, all wages were halved. Mine as well, of course. After I left AEG, I sued the company and received some compensation.

Time and again I heard news about Lawrence's successes with his cyclotrons. Ernst Sommerfeld kept me up-to-date through his father. And there were other machines being developed to increase the voltages which could be generated. For example there was one at the Carnegie Institution in Washington, by Breit, Tuve, Havstad and Dahl [Br28] and another at Princeton University, by Robert J. Van de Graaff [Gr31]. The latter had gone back to an old idea which was to transport electrical charges to an isolated metal sphere by means of suitable strips. He did such a good job that his machines were copied everywhere and they were even produced by industry. I would like to note in this context that Tuve, Hafstad and Dahl were of Norwegian ancestry and that Tuve was a childhood friend and fellow student of Lawrence's.

Then, in 1932, came the first disintegration of an atomic nucleus with artificially accelerated particles. John Cockroft and Ernest Walton achieved this with a cascade-generator which only reached 400,000 volts [Co32]. Incidentally, the principle used for producing the high voltage came from H. Greinacher in Switzerland [Gr21]. Shortly afterwards, Lawrence could confirm the results of Cockroft and Walton using one of his cyclotrons. There was a lot to talk about in Berlin!

However, Hitler was threatening to take power, and I left Germany just in time before it happened. I could already sense that things would not be too good under Hitler and returned to Norway shortly before Christmas 1932.

While I was still working in the 'Dr. Paul Meyer' laboratory I had had an idea for building much better relays. As mentioned before, they had already developed quite an interesting relay long before I got to the laboratory. It could determine the distance to the short-circuit. It was also called a 'distance relay'. However, as already mentioned, it had many faults, was not very precise nor very sensitive. My new idea was much simpler, more robust and promised to be faster and more accurate.

I was pretty well informed about the situation in Norway and knew that the many power stations which were connected together in a so-called 'Samkjöringen' (the 'network') urgently required security distance relays. I also knew that a robust and simple relay would be very useful in Norway. Many electrical companies used only unskilled labour, and complicated, precision engineering was not much use to them.

First of all I selected a relatively small company which I considered suitable for manufacturing my relays. The company was 'N. Jacobsen's Electrical Workshop' (NJEV) in Oslo. I spoke with the director, a Mr. Haug, and convinced him that my relays would be a good thing for him. After short deliberations we came to an agreement, and I was paid 500 Kroners a month, a pretty good salary in those days, and started work at Jacobsen's on April 1, 1933. As I had already completed all the design work for my new relay beforehand I was able to start construction immediately.

I would now like to say more about these relays, although it may only be of interest to readers who are curious about technical matters. Figures 5.1 and 5.2 show such a relay. I used a rod-shaped electromagnet as a voltage sensor which was fed with direct current via a small selenium-rectifier. The magnet's yoke had quite a large pole surface and a fairly strong constriction below. The result was an attractive force on an iron armature, which increased almost linearly with the voltage, even in the lowest region. A bi-metal would then try to pull off the iron armature.

The current-transformer for the bi-metal had a hole at the centre of the iron core which was dimensioned in such a way that the current for the bi-metal increased with the square root of the current, and the temperature rise (and consequently the bi-metal's tension) became proportional to the product of current and time. The interval within which the bi-metal pulled off the armature from the electromagnet was therefore proportional to the ratio between voltage and current, or to the 'impedance' of the short-circuited line. A small hook (a roller bearing) was detached when the armature came off and this activated the high voltage switch placed on the line. It was possible to read the delay-time on a small synchronous watch and thus determine the distance to the short-

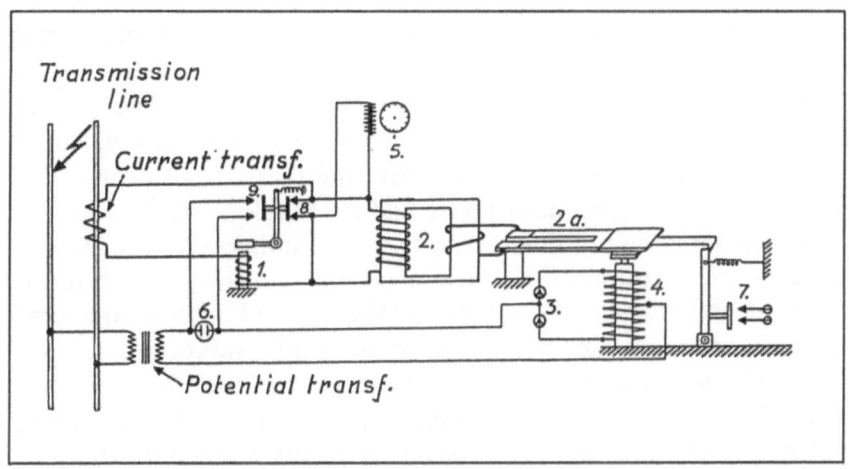

Fig. 5.1: Diagram of Wideröe's relay.

Fig. 5.2: Photograph of Wideröe's relay taken from a N. Jacobsen (Oslo) brochure.

circuit. The relay was cheap to make and did not require any sensitive precision engineering. But it was very accurate.

The shortest reaction-time was only two periods long, i.e. approximately 1/25 of a second. This is of great importance, since, if this time interval (the 'basic time interval') is too long, the generators could get out of phase and the whole system of linked power stations would thus be in danger of collapse. I later wrote a precise description of the 'distance-relay' for the Journal of the Electrotechnical Society of Vienna [Wi37] and I submitted a total of ten Norwegian patents for Jacobsen on this subject.

The relay was completed in the autumn of 1933. I then took my Ford-A on a vacation tour of England, Spain, Italy and Germany, during which I was also going to introduce my relay to the market. To my great regret I found that it was not at all easy to sell the relays. In the end the journey turned into quite an adventure. I met up with my friend Torvald Torgersen in England, and he accompanied me for the rest of the trip. Torvald fell ill on the way. We found out later that he had been infected with typhoid and I caught paratyphoid B as well. We were very lucky to survive those exertions. Torvald is alive and well today, and has a summer house on Skjelöy (near Fredrikstad) close to my sister Else's.

In March 1934, we conducted the first field tests with the relay in Norway, on a line in Vestfold. However, shortly before that, in February 1934, I met my wife-to-be Ragnhild Christiansen in Oslo. I had enrolled at Miss Fearnley's dance academy, in order to learn all the latest dances, and that is where I met Ragnhild, whose parents lived not far from us. We married on November 14, 1934 and spent our honeymoon in Stockholm.

Our three children Unn, Arild and Rolf were all born in Oslo in the years 1936, 1938 and 1941.

Ragnhild occasionally (and unofficially) worked at Jacobsen's during the summer of 1935 and helped me to build and set up the relays. I remember one evening, I was completely absorbed in my calculations and, suddenly realising how late it was, I went into the anteroom where Ragnhild was working and said, "Miss, you may go home now". I had completely forgotten that we were married! And Ragnhild has not forgotten the incident to this day!

In the spring of 1935 we installed the first of about 30 distance relays in the Norwegian power distribution network. Ragnhild and I would often drive around together in my Ford-A and we did almost everything ourselves, from setting up the relays in the factory to installing them in the power stations. The relays have all been very successful and correctly switched off during short-circuits [Wi37]. Quite a few of them are probably functioning today.

I think it was in autumn 1937 when six companies were invited by the 'Samkjöringen' to propose a new network scheme (including protection relays) for the Norwegian power stations. The submitted schemes had to be supported by cost estimates. The six companies were Siemens, AEG, Brown Boveri, the Compagnie des Compteurs, Westinghouse and our little company, Jacobsen. We won hands down. My relays were much faster, much more precise, much stronger and furthermore, they were cheaper than those of the competition.

Then, in 1937, something unusual happened. A gentleman came to see me. His name was Eivind Hansen, he was the director of the large transformer factory 'National Industri' in Drammen and he offered me a job. The factory belonged to the American Westinghouse group which also had an office in Oslo. I was given the impression that I would become Hansen's successor and accepted.

However, I first had to find a successor for my own work at Jacobsen's and then teach him all there was to know about the relays. I found a good man and all went well. Years later the Jacobsen company got into a mess with current limiters or household current-meters. They lost a great deal of money and eventually had to declare bankruptcy.

I spent three years with National Industri, but it was not a happy time. Most of my work consisted of selling Westinghouse transformers and high voltage protection devices (a type of 'Thyrit-

55

Fig. 5.3: Ragnhild Wideröe in the 1930s.

Fig. 5.4: Rolf Wideröe in the 1930s.

protection' against over-voltages, travelling waves and similar things in power lines).

While I was with Jacobsen's I published eight papers, with National Industri not a single one, except perhaps the write-up of one longish lecture on relays which I gave during a Nordic conference of engineers in Copenhagen in 1937. That was typical. With National Industri I was on ice, practically dead. Of course I gave a few lectures on high voltage protection devices, but that was nothing special.

Then, in September 1939, War began. Because of the distance-relays, I had had some contact with 'Norsk Elektrisk og Brown Boveri' (NEBB). This company employed an engineer named Styff who died during the first days of the War, and I suppose that NEBB's director Solberg spoke with Eivind Hansen about me, because shortly afterwards Solberg offered me a position and I replaced Styff in June 1940.

As Finn Aaserud and Jan Vaagen later told me, Styff had been present during my 1937 relay lecture in Copenhagen, so he was aware of my interests. This had been shortly after I had started at National Industri. Finn Aaserud also told me that Niels Bohr gave the introductory lecture, which I certainly must have heard, but cannot remember at all. Bohr's lectures were often a little difficult to understand. A tour of the Bohr-Institute was laid on following his lecture and I definitely wasn't present then. Being in Copenhagen with my wife, I guess I had other things to do – we were probably sight-seeing.

At the time I was close to the 'Physics Association' which was founded in autumn 1938 by students, university lecturers and other interested parties in Oslo. I had managed to persuade National Industri to give financial support to the association, I think they donated about 5,000 Kroners. However, the association experienced financial problems during the War.

According to my friend Olav Netteland's reliable memory, the association was also given a few hundred Kroners to start a magazine. The first and very modest edition was produced in the summer of 1939. It was called 'Fra Fysikkens Verden' (in English: 'The World of Physics') and it still exists today. Until 1956 it was edited by the theoretician Egil Hylleraas, professor at Oslo University. The 54th year's issues appeared in 1992, although by then it had naturally become a somewhat more sophisticated production. I still subscribe to it. When we started the magazine we had counted on financing it with advertising which the printer had arranged for us. Regrettably the clients did not pay and for a while there was no money left with which to continue. In the end however, we managed somehow.

6 Induction from Illinois

It was in the Physics Association in Oslo that something happened which was of great importance to me. In the autumn of 1941 a lecture on particle accelerators was given at the Association and among the work described was that of Donald W. Kerst and R. Serber which had just been published in the American 'Physical Review' [Ke41a][Ke41b]. The lecture was given by the physicist Roald Tangen from Trondheim who became a professor there in 1948 and then in Oslo in 1952.

Kerst described in his article how he had built and put into operation a 'ray-transformer' for electrons which he called 'induction accelerator'. At the end of the acceleration the electrons had an energy which could ordinarily only be achieved by a high voltage of 2.3 million volts. The small piece of equipment had a circular tube with a radius of only 7.5 cm. Probably the most impressive result was contained in the summary: under optimum conditions, the electrons could produce X-rays which corresponded to those emitted by about one gram of radium. If one takes into consideration that one gram of radium had a value of about one million Kroner at that time, it is easy to understand why Kerst's little machine caused such a stir. Its use in hospitals, especially for radiation therapy immediately suggested itself.

Kerst had designed and built the 2.3 MeV ray-transformer at the University of Illinois where he worked. General Electric Company was very interested in his work. They had built the glass ring for Kerst's machine in their Valve Department, exactly to his specifications. When the article was published in Physical Review, Kerst was already on leave of absence at the G.E. Company's Research Laboratory. Here he built further machines of this type.

In a second publication which appeared in the same issue of the Physical Review, Kerst and Serber had formulated a theory of the ray-transformer which, in principle, can be regarded as a natural

Box 4

Roald Tangen, Kerst und Wideröe

Professor Roald Tangen (Oslo University) reports [Ta93]:

"I can well remember the events of 1941. At the time I was working on a small Van-de-Graaff generator which we had built at the Physical Institute of Trondheim Polytechnic. In the autumn of 1941 the Physics Association invited me to give a lecture on modern accelerators in Oslo.

We had been denied access to American magazines by then, and we were completely ignorant of the betatron. A few days before my trip to Oslo a single copy of the Physical Review arrived in Trondheim by ordinary mail. Mysteriously, it had found its way to us. It contained an article by Donald Kerst on the first working betatron [Ke41a]. This fitted well in my lecture in which I went on to explain that Kerst mentioned a German doctorate thesis by a R. Wideröe in which a fundamental equation for the betatron was developed. I didn't know anyone by the name of Wideröe at the time, but I told my audience that the name indicated that he could be a Norwegian. As we were to discover soon enough, Rolf Wideröe was sitting in the auditorium! After my lecture we chatted about this strange coincidence.

42 years went by before we met again. In the same auditorium in which I had spoken about Kerst's betatron, Wideröe, on the invitation of Oslo University, gave an account of his scientific life in 1983. My task was to thank him for his lecture. And while I was at it, I promptly mentioned what had occurred on the very same spot in 1941."

continuation of my ideas of 1928 [Wi28] as well as those of Ernest Walton [Wa29] which we had developed independently and at almost the same time. I shall say more about Walton's important contributions later on.

So it became clear that the ray-transformer did work after all - if things were done correctly. And this was like a thunderbolt for me!

I immediately went back to my calculations for the ray-transformer. For several months I worked on this in parallel with my work for NEBB, and in September 1942 I sent a fairly long paper to the magazine 'Archiv für Elektrotechnik' in Berlin. In it I discussed Kerst's results as well as a few of my latest calculations and formulas [Wi42]. It was published in 1943. Later on I wrote a second article which contained a somewhat adventurous proposal for a 200 MeV betatron. It was submitted to the same magazine in July 1943 but for several reasons was never printed.

A very strange thing happened when my first article appeared. One day, it must have been in March or April 1943, several German Air Force officers came to NEBB wanting to speak with me. Norway had been under occupation since April 1940. I can't remember exactly whether there were two or three of them. I was standing next to my bicycle because I always cycled to NEBB. They asked whether we could go to the Grand Hotel together to talk about something. I countered that it was possible, but first I would have to fix my bike.

In the Grand Hotel they asked me to return to Berlin with them. They said that it could be a matter of some importance to my brother. My brother Viggo, as I already explained, was the director of 'Wideröes Flyveselskap', the airline he had founded. It was closed down because of the War. But my brother had links to people who were trying to get refugees into England and this was, of course, strictly prohibited. They were found out. My brother was arrested, was tried in Oslo and, luckily, was not sentenced to death (as others were). Instead he was sentenced to ten years of severe imprisonment in Germany.

The German officers hinted that it may be possible to release my brother if I helped them. This decided things for me, and I agreed to go to Berlin. Two days later I was flown there for a short visit, and they told me about their plans to build betatrons. If I agreed to help them, they would in turn do everything they could to secure Viggo's release. At the time I knew that he was in Rendsburg jail and that he was not at all well.

They didn't tell me what the Air Force wanted with a betatron; I didn't find that out until later. In any case, I did not know at the time that anyone would want to use betatrons as weapons. I also would not have believed it within the realms of the possible. Of course they had one strong argument: They wanted to catch up with the Americans, regardless of any use the betatron may later be put to. The official line was that all this was being done to develop new and better X-ray apparatus for medicine and for non-destructive testing of materials. Betatrons were small and relatively manageable machines which could replace the high voltage set ups normally required to produce X-rays. They would for instance be useful in field hospitals.

So I agreed to go to Hamburg or, to be more precise, I was 'subjected to compulsory work' with my more or less voluntary agreement (and obviously that of my employers NEBB). Initially I was to develop and build a relatively small betatron for 15 MeV and then perhaps a larger one somewhere near Mannheim. That was in the spring of 1943. I prepared the design of the 15 MeV machine in Oslo until the summer and also planned a few things for further development of this type of apparatus. In July 1943 I also applied for a first patent on betatron-construction which dealt with some details of the injection system.

Although I hadn't been directly involved in particle accelerators since my Aachen days, I had carefully followed the progress made in this field. I had thoroughly re-examined the literature on ray-transformers (later called betatrons) for my 'Archiv für Elektrotechnik' report, and in doing so had come across a series of publications and patents. I still find all this developmental and pioneering work very interesting and would therefore like to mention some of it here, without getting too bogged down in technical details.

In 1937 I had already, quite accidentally, discovered an American patent which introduced a very similar idea to that behind my Karlsruhe ray-transformer. It belonged to Joseph Slepian, who worked for Westinghouse Company. He submitted the patent in America on April 1, 1922, and it was granted in 1927 [Sl22]. Slepian also made use of the induced electric field which appears around the core of a transformer to accelerate electrons within a

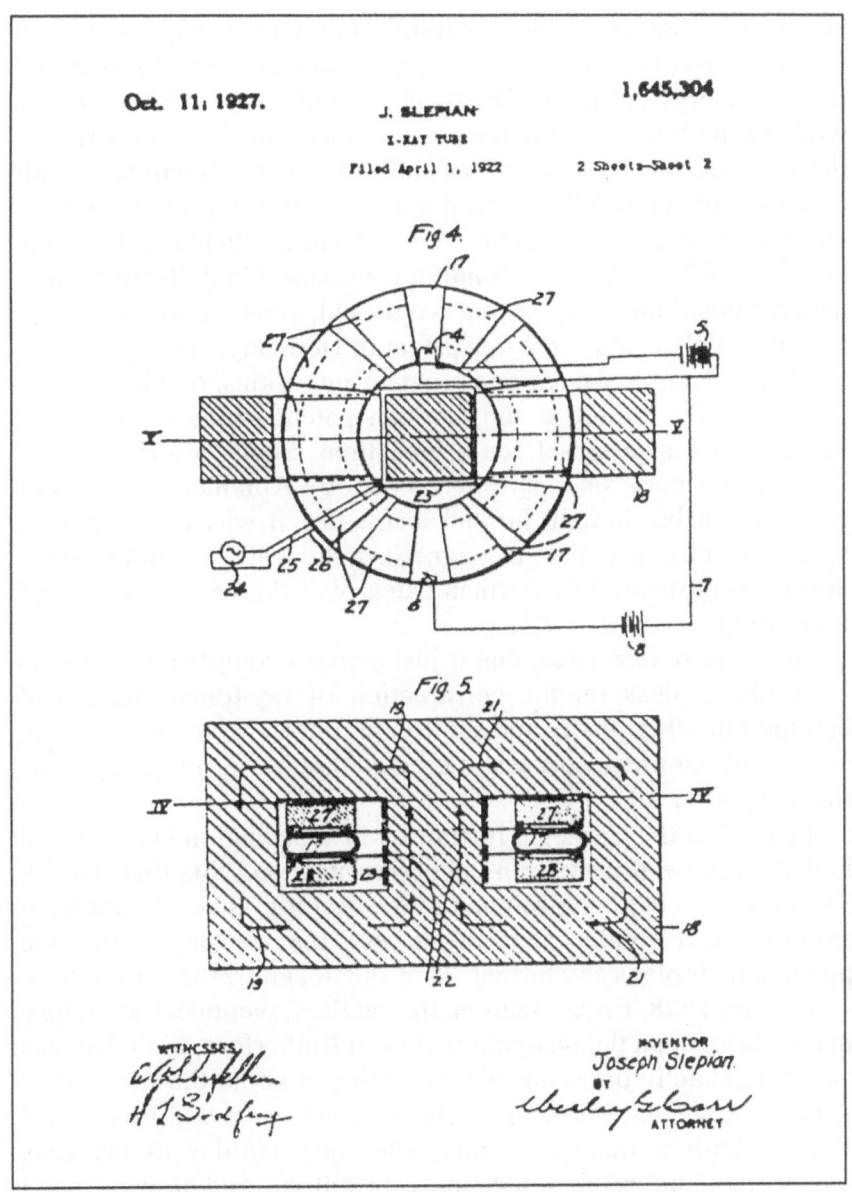

Fig. 6.1: Diagram from Slepian's betatron patent [Sl22].

small disk shaped vacuum chamber. Permanent magnets forced the electrons onto spiral orbits whose radii increased in size as the particle energy mounted. Finally the particles would hit either the walls of the tube or a suitable piece of material. The small size of the machine meant that it was only of use for relatively small energies (under 100,000 eV). It was designed to produce X-rays and was also given the name, 'X-Ray Tube'. Slepian's machine could not follow my '2:1-condition' because his deflecting magnets provided him only with a fixed field, constant in time.

Slepian also submitted this patent in Germany, shortly after he had handed it in to the American patent authorities. As I learnt later [Ka47], a Dr. Smidt at the German patent office doubted the validity of his proposal for a long time. Smidt referred to the famous text book on electromagnetism by Abraham and Becker from which he thought he had learnt that it was impossible to deflect electrons within a magnetic field while simultaneously accelerating them. The German patent was therefore not granted until 1928.

But even before 1928, that is just before I completed my thesis in Aachen, ideas on the construction of ray-transformers had emerged in other places and some experiments were even started, albeit without much success. It seemed that it was not going to be that easy after all.

In 1927 at the Carnegie Institution in Washington DC, a group had already been performing experiments along this line [Br27]. The authors, Gregory Breit and Merle Antony Tuve, should have stood quite a good chance of success, but appear not to have pursued their plans any further. They did not know the 2:1 relation.

Around 1928, Ernest Walton, then at the Cavendish Laboratory in Cambridge, on the instigation of Lord Rutherford, did what was basically exactly the same as I was doing at about the same time, without, however, knowing anything about my work. First of all Walton built a machine which was very similar to my ray-transformer, although much more primitive, and it was not a success. He published his results in October 1929 and included the conclusions of some very interesting and important theoretical deductions on the stability of the electron-orbits [Wa29].

Since Walton's experiments with a ray-transformer were not successful, he built a linear accelerator which was also very similar to mine in Aachen, but again it was much more primitive. It was also fitted with a spark generator to produce the required high voltage. This machine had no chance of success.

And then Walton gave up these investigations. Together with John Cockroft, and with Rutherford's steady encouragement, they built their famous cascade generator, which came to be known as the 'Cockroft-Walton' device. With this he achieved the first nuclear disintegrations with artificially accelerated particles, which I have already mentioned [Co32]. Cockroft and Walton were awarded the Nobel Prize for their achievements in 1951.

Ernest Walton subsequently returned to Ireland where he took up a professorial post. I sent him a letter once, enclosing copy of a lecture I had given at the '5th Nordic Meeting' held in 1983 in Geilo. He thanked me and wrote that he had been able to find much in this lecture that he had not known before. He had heard a little about my work by then.

Walton's work was continued by Leo Szilard and J. L. Tuck at the Clarendon Laboratory in Oxford. They built an iron-free betatron for higher frequencies and this work was also unsuccessful. It is possible to produce suitable magnetic fields for a betatron without using an iron yoke, but this was not made to work until many years later. Leo Szilard, whom I had met in Berlin, had emigrated to England after Hitler had come to power in Germany.

There was another article on ray-transformers in the 'Archiv für Elektrotechnik' magazine published in 1936. It was written by W. W. Jassinski [Ja36] and contained a comprehensive mathematical investigation as well as some technical proposals which I did not find particularly useful at the time.

While I was correcting the proofs of my article for the 'Archiv für Elektrotechnik' in 1943, the physicist Max Steenbeck from the Siemens Company in Berlin published an article in the magazine

'Naturwissenschaften' [St43]. He stated that he had been success-ful in accelerating electrons to approximately 1.8 MeV with a betatron-tube in 1934 and 1935, and that he had also applied for several patents on this matter. I included this as a footnote in my article, on page 545. I also mentioned that Steenbeck roughly described a condition which the magnetic field had to satisfy in order to obtain stable electron orbits in a betatron. This condition was also included in two German patents [Ru33] and [St35] submitted in 1933 and 1935, and in an American patent [St36] which, incidentally, should have been known to Kerst. Steenbeck should therefore be considered as the inventor of this stability condition.

I met Max Steenbeck much later, during an International Conference on Betatrons in Jena in June 1964. It was a pleasant meeting and we had lots to discuss. While I was there, I also gave a lecture on the first ten years of multiple acceleration. The complete text was subsequently published in the periodical of the Friedrich-Schiller University in Jena [Wi64].

However, Steenbeck's stability condition should be regarded as an approximation of Walton's earlier, more general formula-tions. Steenbeck's condition is valid only in the immediate prox-imity of the particle's nominal orbit, whereas Walton's formulas

---

Box 5

The Many Names of the Betatron

Steenbeck and Gund (see also Box 9) called their accelerator machines 'ELECTRON-CENTRIFUGE', while Schmellenmeier and Gans used the name 'RHEOTRON'. Widerøe had introduced the very apt appellation 'RAY-TRANSFORMER'.

In Slepian's patent of 1922 the machine is modestly described as an 'X-RAY TUBE'. Kerst and Serber used the expression 'INDUCTION-ACCELERATOR' in their famous papers of 1941.

It wasn't until 1942 that Kerst introduced the now generally accepted term, 'BETATRON'.

Box 6

About Max Steenbeck

In an interesting book, Max Steenbeck gives an account of his life [St77]. He was born in 1904 and studied physics in Kiel. From 1927 until the end of the War, he worked in the research department of the Siemens-Schuckert factory in Berlin-Siemensstadt where he also completed his dissertation. This is where he conducted his early betatron experiments.

After the War, Steenbeck went to Moscow where he worked for eleven years, mainly on the separation of isotopes. He returned a committed communist, became a professor in Jena, and worked on cosmic magnetic fields, plasma and solid state physics amongst other things, but his main concern was with nuclear energy. He had a good reputation in the GDR as a physicist and held important positions. He later became rather critical of his past at Siemens.

As Steenbeck quotes in his book, he had already developed the basic ideas for a cyclotron and even a first outline for a synchrocyclotron by 1927/28. On the urging of his colleagues at Siemens he then wrote an article for 'Naturwissenschaften' magazine. However, because of a misunderstanding regarding a request for consultation by his superior Dr. Rüdenberg, this article was never published.

Max Steenbeck died in Berlin in 1981.

also apply at greater distances. But Steenbeck's condition was easier to understand than Walton's somewhat more complicated and scarcely disseminated theory. Therefore, Steenbeck is generally regarded today as the author of the (simplified) stability condition. In his first patent (1933), Steenbeck formulated the condition rather vaguely: "...the magnetic field which serves to guide the particles, (is) characterised by the magnetic field dropping off from the centre to the sides...". More was not specified on the subject.

If particles are to stay on a constant circular orbit, it is important to ensure that they are guided by suitable forces. Particles which are not on the nominal course are then gently pushed back. Of course, the force which does the 'pushing' also causes the particles

Box 7

The War of the Patents

Max Steenbeck (see Box 6) and his then superior Rüdenberg (who, like Wideröe, developed relays for power stations) had already submitted a patent on the stability of electron orbits in betatrons in 1933 [Ru33]. From the text it is apparent that they were aware of Slepian's patent [Sl22] (in those days, and until after World War II, it was not yet mandatory to make precise references to 'previously performed or published work' when submitting patents).

Siemens subsequently asked Steenbeck to run a secret project to construct such a tube. He had read Wideröe's thesis [Wi28] in the meantime, and thus took into consideration the 2:1 ratio between steering and accelerating magnetic fields. The machine of 1934/35 was able to accelerate electrons to 1.8 MeV. However, the number of accelerated electrons was far less than had been expected, so the work was stopped. But during the course of this investigation Siemens submitted a second patent for Steenbeck in Germany [St35], the USA [St36] and in Austria, in which protection by patent for both the stability condition and the 2:1 ratio, among other things, was specifically applied for. General Electric USA applied to Siemens for a licence to use this patent and, according to Steenbeck, this was granted on December 6, 1941 (shortly before America entered the War).

Kerst (University of Illinois) had already published his first betatron results in October 1940 [Ke40a] and immediately afterwards he submitted the betatron for US-patenting (for General Electric) [Ke40b]. It is very similar to Steenbeck's patent, although clearer in its formulation. Kerst's famous work on the 2.3 MeV betatron followed it in 1941 [Ke41a]. Kerst makes no mention of Steenbeck's patents in these publications (neither does he mention his own), but he does refer to the work of Wideröe and Walton, and later also to Breit, Tube and Jassinski. It is unusual for patents to be mentioned in scientific publications.

Following Kerst's famous paper, Siemens, prompted by Steenbeck, took up the construction of betatrons again and appointed X-ray-engineer Konrad Gund to do this. Siemens were able to assert their rights in 1954 and received compensation from BBC for their alleged use of Steenbeck's patents.

During 1943/44 Wideröe submitted ten patents on the betatron for BBC and there would be more later.

to snake around their intended course, just as a sleigh would when in a rut. These backward and forward movements are called betatron oscillations. They can be radial as well as vertical (on a ring placed in a horizontal plane). Appropriate correcting forces are created in Walton and Steenbeck's magnetic fields within the betatron. They have to decrease proportionally to $1/r^n$ going outwards, whereby the number n lies between 0 and 1. This is a more precise formulation of the intention in Steenbeck's patent.

In 1946, Donald Kerst provided a precise illustration of the history behind the betatron in a comprehensive and very interesting article for Nature magazine [Ke46]. He expounded on all published and unpublished work on the subject, in as far as he was aware of its existence. It seems certain to me however, that the basic idea for the construction of a betatron or ray-transformer was developed independently and in different places at the same time.

By the end of the summer of 1943, while still in Oslo, I had progressed far enough with my ideas and preliminary studies to be able to start constructing a betatron.

However, from July 25 to August 3, 1943, 'Operation Gomorrah' had been executed: during the course of five air attacks Hamburg's centre and some outlying areas were almost completely destroyed by American and British bombs. An enormous number of people were killed, probably more than 50,000. After this, Hamburg was regarded as a 'relatively safe' place, because it wasn't thought that anyone would find it worthwhile to subject the city to such intensive bombardment again.

And so I began my work in Hamburg, although I often returned to Oslo which is where I wrote many of my reports. During this period (the second half of 1943), and always with the assistance of Ernst Sommerfeld, I submitted another four patents in Germany which concerned the construction of betatrons as well as a very special patent, which I shall describe later on.

69

7 The Hamburg-Betatron

My wife and our three children remained in Oslo while I started working in Hamburg in August 1943. I was an employee of NEBB during my entire stay in Hamburg and my wife continued to draw my salary in Oslo. So we had no problems – apart from the separation. I had rented a room in a beautiful house in a leafy suburb of Hamburg.

My first, and probably most important, contact in Hamburg was Richard Seifert. Later he became doctor honoris causa of the Universities of Hamburg and Hanover. I can't remember exactly how this contact came about. In any case, he was the owner and director of a medium sized factory, founded by his father in 1892, which had a good reputation internationally. This factory had already begun manufacturing devices for X-rays in 1897, i.e. just two years after Röntgen had discovered these rays, and during my time they manufactured these, mainly for non-destructive testing of materials, such as welded joints. Not only did they manufacture a standard range of products, but they also dealt in clients' special customised orders. During the War they were major suppliers of apparatus for materials-testing for the German air-craft industry.

Seifert was a hard-working and honest man and I had the greatest respect for him. He was very supportive to me in my strange predicament. Years later we would often visit the youngest of his three daughters, Elisabeth, when we passed through Hamburg. By then she had taken over the management of the factory. The various departments were later relocated to Ahrensburg near Hamburg.

Also in Hamburg I had a wonderful collaborator and colleague, the physicist Dr. Rudolf Kollath, who had previously worked in the aluminium factories in Sauda (near Stavanger in Norway) as well as at AEG in Berlin – I believe with Professor Ramsauer. Later on he became a professor in Mainz and he also wrote a very nice book on particle accelerators which appeared in 1955 [Ko55]. The second edition was much more comprehensive. It included contributions by several well known scientists and came out in 1965.

While I was working in Hamburg I wasn't really answerable to anyone in particular. The only person to whom I had some contact of that type was Air-Force Group Captain Friedrich Geist. I occasionally paid him short visits at his office in Berlin. He was a sensible man and not without charm. After the War ended I never heard of him again, except for the following information which was conveyed to me by Jan Vaagen in 1983: Apparently David Irving refers to him quite extensively in one of his books. I had no connections whatsoever with anyone of higher rank with regard to my work.

On the other hand, I did have a lot to do with a relatively small, private company which acted as mediator between those in Berlin financing my work, (which were the Air-Force or the Ministry of Aviation, the 'Reichsluftfahrtministerium' RLM), and myself. The head of this small company was called Hollnack and he was a rather strange and somewhat highly-strung person. I remember he had a high regard for Nietzsche and probably (although we never spoke about it) also for Hitler. Apart from the betatron he appeared to have some other business with aluminium alloys, but this was of no interest to me. He administrated my Hamburg project. Hollnack (his first name was probably Theodor – he changed name after the War [Gi93]) claimed to have very good relations to high-ranking personalities in Berlin, and I suppose he let or negotiated contracts between the Ministry of Aviation (or other official bodies) and individuals or companies.

I met Hollnack one more time after the War in Waldshut (Germany), after he'd telephoned me. He wanted to claim rights on patents which had come about thanks to his 'mediation' in Hamburg, but of course, that was not possible. All the patents I submitted at that time belonged to Brown Boveri Company BBC in Baden (as NEBB was a subsidiary of BBC) for whom I had already worked in Oslo.

71

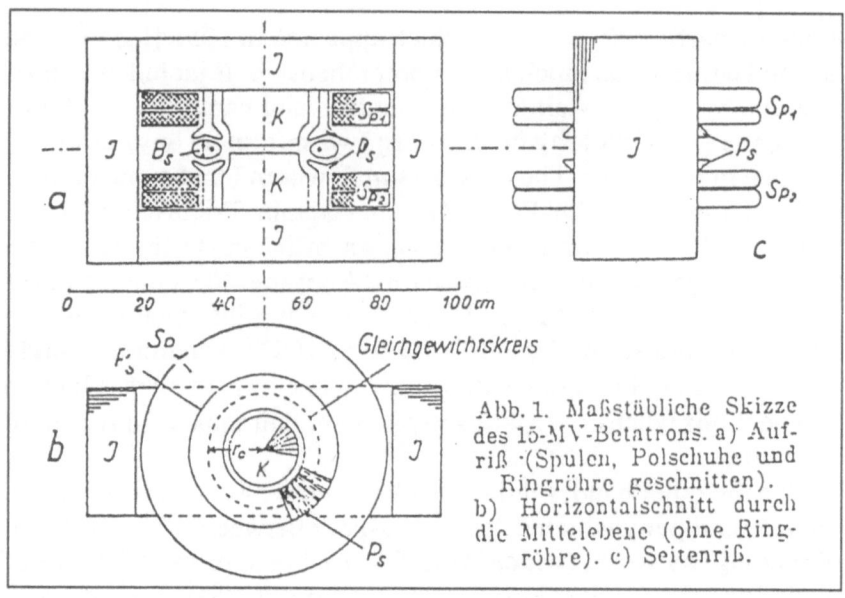

Abb. 1. Maßstäbliche Skizze
des 15-MV-Betatrons. a) Auf-
riß ·(Spulen, Polschuhe und
Ringröhre geschnitten).
b) Horizontalschnitt durch
die Mittelebene (ohne Ringröhre). c) Seitenriß.

Fig. 7.1: Diagram of the Hamburg betatron [Ko47].

Fig. 7.2: Photo-
graph of the
Hamburg
betatron; ETH
Library Zurich
Hs 903: 614.

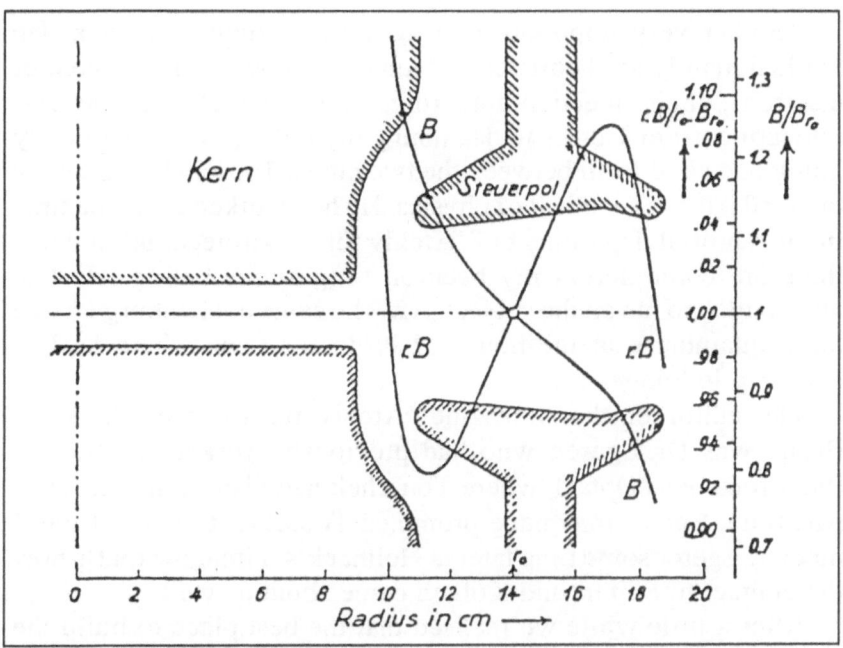

Fig. 7.3: Shape of the pole-pieces of the Hamburg betatron, for meeting the required beam stability condition [Ko47].

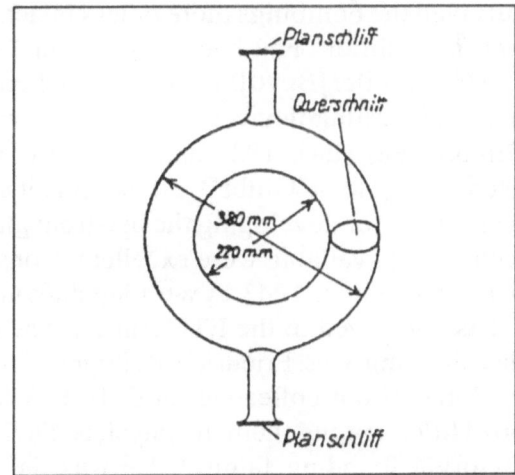

Fig. 7.4: The tube of the Hamburg betatron [Ko47].

Another very important colleague was Bruno Touschek. He worked mainly on theoretical calculations about the movement of electrons, their injection into rings and other effects. He was relocating from Berlin to Hamburg or, rather, was continually going back and forth between the two cities. Touschek was a very talented Austrian student of physics. He had worked for some time in the editorial department of 'Archiv für Elektrotechnik' and had therefore come across my betatron suggestions before. He had also written to me on the subject. After he came to Hamburg I made his acquaintance at the house of Professor Lenz where he had taken up lodgings.

The editor of the Magazine 'Archiv für Elektrotechnik' in Berlin was Dr. Egerer, who had previously worked for 'Löwe' (later renamed 'Opta'), where Touschek had also been employed part-time. Egerer may have prompted Touschek to contact me. I met Dr. Egerer some time later at Hollnack's. I imagine that's how the contact to Seifert and Kollath came about as well.

After a little while we realised that the best place to build the betatron was at the big X-ray-tubes and radio-valves factory called 'C. H. F. Müller', also known locally as 'Röntgenmüller'. Their buildings lay in the north of Hamburg, in Fuhlsbüttel, and had survived the bombings more or less intact. This factory, which was rich in tradition, had been founded in 1865 by the glass-blower C. H. F. Müller [Be90]. They also supplied X-ray-tubes for Seifert's materials testing devices. It has been owned by the Philips group (Eindhoven) since 1927 and still exists under the name 'Philips Medizin Systeme GmbH'. At the time it seemed to be particularly well suited for developing the betatron: glass-blowing and vacuum techniques available were excellent. Construction started in October or November 1943. A working drawing of March 1944 at scale 1:1 is conserved in the ETH Zurich. The engineer responsible for that drawing was Friedrich Reiniger.

Some of our colleagues at C. H. F. Müller were very Nazi and pro-Hitler, among them the physicist Dr. Müller (no relation of the factory's founding family). He was the physicist Walter Müller (born 1905) who had developed the famous counter tube with Hans Geiger in 1928. However, Müller never used his first name to sign documents, only ever 'Dr. Müller'. He was a nice and hardworking man, quite popular, but we were always very careful when we spoke with him. According to a later report by Herman Kaiser [Ka47], this Dr. Müller also submitted or prepared a series of seven patents for the betatron (file references are quoted), but I have no recollection of this. In the ETH Library Archives there is a fifteen page long report by Dr. Müller [Mu43] in which he expounds on the betatron as well as on its theory.

Every now and again I was permitted to leave Hamburg by air for holidays in Norway. The journeys were often a little problematic. Once, I think it was in December 1943, I was trying to get home for Christmas. We had to wait a long time in Denmark because of fog, but we arrived in Oslo just in time for the celebrations.

I eventually found out why the German Air Force was so interested in the betatron. Physicist Dr. Schiebold from Leipzig, a specialist on non-destructive testing of materials using X-rays among other methods (after the War he became professor in Magdeburg) had had the idea that it would be possible to build an X-ray tube with a concave cathode, a bit like a concave mirror. The electrons would then be focused on the anode and this would cause the X-rays to be emitted in a narrow bundle. With sufficiently high voltage it would then be possible to achieve high radiation intensities at long distances.

Thus it may even be possible to kill the pilots of enemy aircraft, or detonate their bombs. This was one of the 'death rays'. With the 'wonder weapons' of Peenemünde, the 'death rays' had become an urgent necessity for war-time propaganda. At the time, the use of far reaching electromagnetic waves was probably quite conceivable since bomber planes were being precisely guided far over British territory by radio waves, i.e. electromagnetic radiation. The classic example was the night attack on Coventry which would have been inconceivable before.

Box 8

The Mysterious Death Rays

Back in 1935 the use of 'death rays' was a suggestion raised in England as a possible defence against the eventuality of German air attacks. These 'death rays' were intensely focused electromagnetic waves. Their workings were described in almost the same terms as Dr. Schiebold's later suggestion in Germany. This is recounted by a then member of British intelligence, the physicist R. V. Jones, in his book 'Most Secret War' [Jo78]. The British soon rejected this proposal because it was out of the range of the available technology. The SDI projects in the USA are a modern version of these ideas.

Schmellenmeier's contribution in Edgar Swinne's book 'Richard Gans' [Sw92], recalls wartime calculations Professor Gans performed for the Rheotron (see Box 9). Gans arrived at the conclusion that if X-rays were very 'hard' (about 100 million volts) they would no longer be emitted in all directions (as they would at low voltages), and instead they would be tightly bundled. That was a completely new idea at the time, since installations of such high voltage did not yet exist. However, Gans had also noted that he had forgotten to include the 'Compton effect' in the calculation and that the whole procedure was therefore in fact impossible. Despite this, Schmellenmeier indicated in a report justifying the continued building of the Rheotron that, "aeroplane engines could be 'pre-ionized' with the bundled, highly penetrating radiation, so that the ignition would fail, the machines could no longer fly and would thus enter into the flak zone".

However, the main reason behind this statement was to continue the Rheotron project in order to save the life of Richard Gans who was of Jewish origin or, as the terminology of the time would have it, a 'privileged non-Aryan' – and Schmellenmeier finally did achieve this.

The fact that betatrons could reach relatively high electron energies and that this could be used to make stronger bundles of X-rays must have given the death-ray advocates renewed hope. And that is how some projects (such as Wideröe's) were financed by the German Aviation Ministry. Others however, such as Gund's and Schmellenmeier's, were financed by the German Research Authority.

It appears that Dr. Schiebold hawked his ideas about. He spoke to physicists who must have thought him a hopeless case, but he also tackled some influential people in official capacities who were not in a position to make informed judgements. Most people probably dismissed him as a harmless lunatic, but some must have been convinced because the Air Force, i.e. the German Aviation Ministry (RLM) and Command of the Luftwaffe, provided a certain amount of support for his 'death ray'.

In order to conduct some test experiments for this 'death ray', a still unused and unpacked X-ray apparatus with a high voltage supply of a little over one million volts (made by means of a sort of cascade circuit), was taken from a hospital in Hamburg to a small military airport called Groß-Ostheim (today 'Großostheim') in the region of Hanau. If I remember rightly, Richard Seifert organized this tests and Hollnack was their administrator. However, both engineers and technicians quickly understood that the danger to themselves operating the machine on the ground was far greater than to the pilots and bombs in the enemy aircraft.

Still, a ray-transformer or betatron could produce X-rays of many million volts and in doing so one could, in principle (purely on the grounds of the laws of physics), improve the 'bundling' of the beam with an increase of energy. To a certain extent, the effective range could be increased. This seemed to be the reason for the German Air Force's interest in the betatron. I wasn't really supposed to know anything about it, and we only ever talked about the betatron in terms of its importance to medicine. As it turned out this was actually correct.

By November 1943 I had developed a three-phase plan [Wi43c] which provided first for the construction of a 15 MeV betatron in Hamburg, then a 200 MeV betatron and finally an experimental station in Groß-Ostheim for even larger installations. Everything apart from the first phase obviously remained an illusion.

Our work in Hamburg soon confirmed that the step from Kerst's 2.3 MeV machine (USA) to our planned 15 MeV ray-transformer was the right one. Of course, all we wanted in principle was to achieve as much energy as possible, but at 15 MeV we did not expect any imminent problems with the iron yoke (which was very similar to that of an ordinary transformer). However, these problems did appear when we built the first 31 MeV machine for Brown Boveri in Baden, as I shall explain later.

On one occasion I went to Rendsburg to visit my brother in prison. He was not at all well, he was suffering from some ailment, but I don't know what it was; most probably it was a manifestation of malnutrition, but it could have been pneumonia. I tried to cheer him up and to get him better treatment, but it turned out that the people with whom I was in contact did not have sufficient influence to have him released. Perhaps they did what they could, but it was not enough. Viggo was given slightly better food and later transferred to a penal colony near Darmstadt where he was permitted to work out of doors, chopping wood in the forest and digging over the soil in the garden, and I am sure that this helped him a lot. At the end of the War he was liberated from this camp by the Americans.

Fig. 7.5: Wolfgang Paul (left) and Rolf Wideröe at the 1992 Accelerator Conference in Hamburg. Photograph by Pedro Waloschek.

─────────── Kasten 9 ───────────
Betatrons in Germany

The late Professor Wolfgang Paul (University of Bonn), a pioneer of particle physics and high energy accelerators in Germany, who was awarded the Nobel Prize in 1989 for the development of the 'ion-trap', described Germany's War-time betatron projects in 1947 [Pa47]. He mentioned Wideröe's and Steenbeck's work and the developments subsequent to 1941:

"Kerst's success meant that work on betatrons, as Kerst later named his machine, was also resumed in Germany. The construction of such an electron accelerator was approached from a total of four different directions. At the forefront of the betatron builders' intentions was the exploitation of fast electrons or their X-rays for medical-therapeutic purposes and for testing materials. Use of the betatron as a research instrument for physics was considered of secondary importance. There were projects by K. GUND at the Siemens-Reiniger-Werke, Erlangen, prompted by STEENBECK for machines of 6 and 25 MeV; further by WIDERÖE for 15, 100 and 200 MeV, by BOTHE and DÄNZER for 10 MeV and by GANS and SCHMELLENMEIER for 1.5 MeV. Of these, GUND's for 6 MeV and WIDERÖE's for 15 MeV were completed by 1945, whereas the others did not progress beyond planning or only went as far as the construction of the magnet-systems."

Paul then goes on to describe the two successful machines by Gund and Wideröe in detail and to compare them.

As Professor Paul further reports [Pa93], he and his tutor Hans Kopfermann also wanted to build a betatron in Göttingen. However, when they heard about Gund's project they offered their assistance to Siemens and were able to conduct first experiments with the 6 MeV betatron (then 5 MeV) in the spring of 1944 in Erlangen. During the American occupation in 1945, an order to dismantle the betatron had been issued, but Paul and Kopfermann succeeded in preventing this with the help of the British Military Government. In 1947 they were able to transfer the betatron to Göttingen where they and others used it for several nice experiments [Gu50]. They also succeeded in extracting the electron beam [Gu49].

This betatron has been an exhibit of the Smithsonian Institution Museum in Washington since the 1960s.

Working in Hamburg was not always without complications. We often had to flee to the basement during air attacks and would have to wait there until the danger had passed. When we came up again there was always a big question as to whether the betatron-tube was still sealed and sufficiently evacuated. However, my sojourns in the cellar also had their up-sides. Down there it was possible to think in peace about possible improvements and to let the imagination run free. This is where I thought up my 'lens-road', a precursor of the 'strong focusing' for particle accelerators which was introduced some time later. I also submitted these ideas for patenting, always helped by my friend Dr. Ernst Sommerfeld who took care of everything in Berlin.

Therefore the War and the limitations of our 15 MeV betatron gave me the opportunity to spend a lot of time in meditating about improved steering and focusing of particles in circular accelerators and in thinking up other new ideas.

8   The Invention of the Storage Ring

At this point I would like to recall an important event of my Hamburg period. It happened during the autumn of 1943, on one of my vacation trips back to Norway. Ragnhild and I were staying in a hotel in a forest in Tuddal, near Telemarken, and Ragnhild unfortunately took rather ill with pneumonia.

I was lying on a grassy hill one day, observing the clouds in the sky, when I noticed two clouds moving towards each other, as if they were about to collide. This started me thinking about cars in frontal collisions and inspired me to make the following consideration: On frontal impact, most of the kinetic energy of both cars is transformed into destructive energy. On the other hand, if a car collides against one which is at rest, only part of the kinetic energy contributes to the destruction. Quite a considerable amount of it is used up to hurl the previously stationary car away and therefore is not available to destroy the two cars. This is a result of the laws of mechanics.

I had thus come upon a simple method for improving the exploitation of particle energies available in accelerators for nuclear reactions. As with the cars, when a target particle (at rest) is bombarded, a considerable portion of the kinetic energy is used to hurl it (or the reaction products) away. Only a relatively small portion of the accelerated particle's energy is used to actually split or destroy the colliding particles. However, when the collision is frontal, most of the available kinetic energy can be exploited. For nuclear particles, relativistic mechanics must be applied, and this would cause the effect to be even greater.

However, it is not so easy to achieve head-on collisions of very small particles against each other. A large number of particles are required and they have to be tightly bundled in order for any two to stand any chance of ever colliding with each other. At the time, I was thinking of atomic nuclei. Since Rutherford's experiments their approximate size was known and I could therefore estimate the probability of a collision. However, given the particle beams available at that time this was an utterly hopeless venture.

And this is where I had my second idea. If it were possible to store the particles in rings for longer periods, and if these 'stored' particles were made to run in opposite directions, the result would be one opportunity for collision at each revolution. Because the accelerated particles would move very quickly they would make many thousand revolutions per second and one could expect to obtain a collision rate that would be sufficient for many interesting experiments. I gave the name 'nuclear mill' to this storage ring, or rings, in which the collisions were to take place.

This exceedingly simple principle was not conceived of again until 1956, i.e. thirteen years later, in the USA [Ke56] [O'N56], when it was developed further and eventually put into practice. Also, in the USSR, at Novosibirsk, similar ideas appeared. However, the first storage ring was put into operation in 1961, and it was not in either the USA or the USSR, but in Italy. Many storage rings used in high energy physics were built in accordance with the principle of this first Italian machine.

After I returned to Hamburg I spoke with Touschek about my ideas and he said that they were obvious, the type of thing that most people would learn at school (he even said 'primary school') and that such an idea could not be published or patented. That was fine, but I still wanted to be assured of the priority of this idea, and I thought the best way to do this would be to submit a patent. I telephoned my friend Ernst Sommerfeld in Berlin and we turned it into a very nice and quite useable patent which we submitted on September 8, 1943 (see facsimile in Appendix 1). This was given the status of a 'secret patent'. It was not until 1953 that it was retrospectively recognized and published [Wi43a].

But we had taken Touschek's objections into consideration and did not state anything about the favourable balance of energy during a frontal collision in the patent, as this was considered a well known fact. Even so, Touschek was pretty offended.

However, the time was not yet ripe in 1943 for constructing storage rings. It was only years later that the accelerator experts came to be in a position to propose and build realistic storage rings for physics experiments. Before that, a whole series of technical problems had to be solved. It was even necessary to develop entirely new technologies. BBC therefore earned nothing from this patent.

The only particle accelerators which I was in a position at that time to propose for my 'nuclear mill', were the betatrons so successfully built by Kerst. And those weren't really suitable for anything other than electrons. But I suspected that soon there would be ring-accelerators for other particles, quite apart from the cyclotrons already in existence. The latter, however, could not be used as storage rings because they did not have stable particle orbits with a constant radius.

The first accelerator (apart from the betatron) in which the particles turned around in a stable orbit was the 'synchrotron', developed in several places after 1945. I worked on this subject myself. I shall later come to describe the problems related to these machines. It was not until ten years later, i.e. in 1956, that this new type of accelerator was proposed as a storage ring, by then a straight forward and natural idea.

I was not worried about the missing technology for my patent to be realized as yet. My main concern was the principle – for which I wished to secure priority for myself. So I put the vacuum problem (and others) to one side for the time being, although this had already caused me difficulties with Professor Gaede when I proposed my first ray-transformer in Karlsruhe. Now this was obviously a completely unresolved problem, because an even better vacuum was required to store particles for a longer time without colliding with molecules of the residual gas.

I was also aware of the fact that the orbits lacked stability. I had been dealing with this problem since my time in Aachen and knew how difficult it was. Kerst was the first to solve it in practice. There was also the problem of getting equally charged particles to run in opposite directions within the same tube. I came up with an adventurous proposal by which the particles would be guided by electrical fields. This never became realisable. It was simpler to use two rings with steering magnetic fields, and indeed this is how it was later done.

But none of this changed the fact that the best way to exploit the accelerated particles' energy was by frontal collision – today this is known as a 'collider', of which there is a variety of types – and that storage rings provided the particles with more opportunities to collide (in fact many thousand per second), as is explained in my patent. Bruno Touschek could not shatter my optimism. Years afterwards he pioneered work in this direction himself – but more of that later.

In the meantime, construction of the 15 MeV betatron went according to schedule and it began operations in the summer of 1944. Intensity was very low at first, but eventually it could be increased sufficiently to be comparable with Kerst's second betatron of 1942 [Ke42]. This betatron accelerated electrons up to 20 MeV. Given his higher frequency (180 Hz, i.e. 3.6 times as much as we had) and his greater electron injection voltage (20 KV as opposed to our 7.5 KV), Kerst did in fact achieve approximately thirteen times our maximum intensity.

Later on, the X-rays produced did on occasion correspond to the radiation equivalent to an entire kilogram of radium, as it was reported by Kaiser [Ka47] – but usually it was only equivalent to about 30 gram, which is already pretty dangerous.

In the beginning we used a hot cathode to produce electrons but the filament could only provide us with the greatest intensity when it was in a favourable position. The result was that the intensity was constantly changing. Rudolf Kollath called it our 'squirrel'. Later we used an oxide cathode and the source became more stable and the intensity more constant.

As I mentioned earlier, a 6 MeV betatron was built at around the same time in the Siemens-Reiniger factory near Erlangen, following Max Steenbeck's proposal. The X-ray specialist Konrad Gund was appointed to the job at the end of 1941. I went to visit him in November 1944. For a variety of reasons I did not believe that the machine would ever work. In particular, there were problems with the vacuum tube made of ceramic material, which is a very good insulator. Electrons leaving their nominal orbit penetrated the walls where they accumulated and eventually caused disruptive discharges in the wall which caused the vacuum to collapse. I was able to prevent this effect in my machines by using glass of weak conductivity (boron-silicate glass, C9) for the tube.

But we also discussed the machines' frequencies and I think I managed to persuade the Siemens team to use 50 Hz rather than the higher frequency of 550 Hz used by Gund. I heard later that this betatron was taken to Göttingen after the end of the War. Konrad Gund collaborated successfully with physicists and he went on to attain a doctorate [Gu46]. Gund, however, was psychologically unstable and he and his wife committed suicide in 1953.

One day, we were visited in Hamburg by Professor Gentner from Heidelberg and Professor Kulenkampp from Tübingen. They were full of praise for our results.

By the autumn of 1944 our betatron had progressed sufficiently for me to be able to hand over the work to Dr. Kollath and Gerhard Schumann. They did a good job and later published an in-depth report on it in 'Zeitschrift für Naturforschung' [Ko47].

At about that time I was invited to a meeting at the Kaiser-Wilhelm Institute in Berlin at which were present several physicists. The meeting took place in a beautiful garden. I think Heisenberg may have organized it, but perhaps it was Gerlach. This conference was of a purely scientific nature. We all spoke freely and said exactly what we meant. As there weren't any Gestapo men present, nothing was kept secret.

We unanimously agreed that Schiebold's fantasies should be called off as they were so utterly unrealistic. On the other hand, it was decided that the betatron was a very interesting machine, especially with regard to the medicine and nuclear physics of the future. The hopeless 'secret project' which aimed to shoot down aeroplanes with X-rays produced by betatrons was dropped in its entirety. The development of betatrons however, was to continue. Of course, in this case it was possible to maintain the official justification that the betatron project was of importance to medicine. It did not cost a lot of money and in any case, money did not play a tremendous role in Germany at that time.

I had several meetings with the directors and design engineers of Brown Boveri & Co. (BBC) on the construction of a 200 MeV betatron. Richard Seifert had awarded BBC a preliminary order for planning such a machine (from the German Aviation Ministry [Wi44]). Several possibilities were investigated and detailed drawings were produced, as was later reported by Hermann Kaiser [Ka47], but these plans were never realised. BBC's factories in Mannheim were in a fair state of destruction, and after Germany was occupied, all mention of these plans ceased completely. Kaiser, however, judged these efforts as the most progressive plans for future betatrons in Europe.

After receiving a final payment for my work from Hollnack I returned to Oslo in March 1945. This time I took the train. We had to stop in Denmark several times, because parts of the railway tracks had been sabotaged. I had another stop in Copenhagen to get my documents in order at the Norwegian consulate.

Our betatron was not the only thing that worked well in those days. The British Army was also doing well and rapidly approaching the city of Hamburg. The German Aviation Ministry therefore ordered the betatron to be moved to Kellinghusen, near Wrist, approximately 40 km North of Hamburg in central Holstein, as Dr. Werner Fehr from C. H. F. Müller reports [Fe81]. Here Seifert's family offered the use of a dairy in which Kollath and Schumann could install the betatron.

On May 3, 1945 British troops occupied the centre of Hamburg. They met no resistance. It appears that Hollnack immediately went over to their side with a full show of flags. Germany surrendered unconditionally on May 7, 1945 and we may assume that Berlin ceased to finance the work on the Hamburg betatron from that moment on. However, Kollath and Schumann got the betatron running in Kellinghusen without any major hitches, and even continued working and taking measurements until December 1945, as is documented in notes now kept in the ETH library [Ko45]. Bruno Touschek had also moved to Kellinghusen, at least for a while, as a few of his manuscripts regarding the theory of betatrons are marked with the name of that place [To45].

In 1947 Kollath and Schumann wrote the extensive report mentioned earlier on the performance of the betatron [Ko47], a work which was mostly done in Kellinghusen. In a footnote on the first page of this report are the words, "We would like to thank the gentlemen of the company C. H. F. Müller for their active support at all times and for their commitment to continuing this work". The footnote which immediately follows states, "We would also like to thank Mr. Richard Seifert, factory owner of Hamburg, for his willingness to offer us his assistance at all times."

From this we may deduce that both companies found a way for financing the work in Kellinghusen as they were themselves involved in the construction of X-ray apparatus (and are still today). Later however, the construction of betatrons was taken over by larger companies such as BBC and Philips-Eindhoven.

In December 1945, the British authorities decided to take the betatron, as part of the booty of war, from Kellinghusen to the Woolwich Arsenal near London. Apparently, Rudolf Kollath later on took charge of its operation in Woolwich where it was used for non-destructive X-ray inspection of steel plates and such like. The machine has since disappeared without a trace. Many, including myself, later attempted to find it, but with no success. It was most probably scrapped.

With hindsight, 1943 and 1944 were very positive years for me, despite all the problems. During this period I submitted ten very important patents on the construction of betatrons for BBC. When I returned to Norway in March 1945 I had already started thinking about issues relating to an even better accelerator which today is known as the 'synchrotron'.

I would like to make another mention of my colleagues at C. H. F. Müller in Hamburg. As I recalled earlier on, there were Dr. Rudolf Kollath, Gerhard Schumann and Bruno Touschek. We were also actively supported by the engineers and designers in the company. I would like to make special mention of the head of the laboratory in which I got my working place, Ing. A. Kuntke. He lived outside of Fuhlsbüttel. His home was in a lovely wood and I visited him there a few times. I also remember Dr. Werner Fehr, who was very active at the time and whom I met several times later on in Remscheid. A few years ago he sent me a nice photograph of the Hamburg betatron. He wrote an interesting booklet on the history of C. H. F. Müller. I have already mentioned Ing. Friedrich Reiniger. He is still alive, as is Ing. Gert Krohn who was working on a linear accelerator for industrial purposes – if I remember rightly. We had many interesting discussions and the atmosphere at C. H. F. Müller was pleasant and cooperative.

I met Schumann once more in January 1945 before I left for Oslo and then heard no more of him for a long time. In his CERN-Report 'The Touschek Legacy' [Am81] the famous physicist Edoardo Amaldi writes that Gerhard Schumann (born 1911 in Dresden) studied in Halle and Leipzig (where he worked with Smekal) and went to Heidelberg in 1950 to work with O. Haxel. He later studied fall-out problems by means of filtering methods and became an expert on exchange phenomena in the atmosphere.

During my period in Hamburg I also met other scientists with whom I had a good rapport. Political issues were very rarely mentioned during our conversations. However, I do believe that most Germans knew nothing of Hitler's atrocities against the Jews. We certainly never spoke about it.

Amongst the people I met was Dr. H. Suess. He lived a little way out of Hamburg and worked with O. Haxel and H. J. D. Jensen. Suess later became a professor at the University of California. At that time he was concerned with the abundance of the elements in the universe. Dr. Suess was absolutely opposed to Hitler and I was able to converse quite freely with him. He gave me the impression

Figure 8.1:
The AdA
storage ring
in Frascati.
Photograph
by Peter
Joos
(DESY).

that scientists were doing everything in their power to prevent nuclear bombs being built in Germany. The only potential they perceived in splitting uranium was as a future source of energy. A small reactor was under construction in southern Germany, but one may presume that this was little more than a diversionary tactic.

I don't think that my work in Hamburg was used in any way for purposes of war propaganda, not even by way of a hint, especially not after the Berlin meeting. The miracle weapons were expected to come from Peenemünde. In my opinion, German morale was at a very low ebb during the second half of 1944. Of course the government did attempt to improve the general mood with a few propaganda tricks, and Goebbels was, after all, a talented copy writer. Nevertheless, I imagine that most people didn't think they stood a chance.

Before we take leave of my time in Hamburg I would like to say a few words about Bruno Touschek. He was a very young man at the time, a student. Touschek's mother was Jewish, and, obviously, that caused him many difficulties. As I mentioned before, Touschek worked with Dr. Egerer at 'Opta' (previously 'Löwe') in Berlin. Dr. Egerer was also the editor-in-chief of 'Archiv für Elektrotechnik' at that time. It was probably Egerer who brought Touschek to us in Hamburg.

Touschek lived in Professor Lenz' house, where, as I said, I first met him. Lenz had some psychological problems and whenever there was an air-raid he would be so scared, that Touschek had to carry him into the cellar. Touschek was able to listen to Lenz and Jensen's lectures at the University while in Hamburg, but he was not officially registered as a student. As a 'non-Aryan' he had already been forced to stop his physics studies in Vienna.

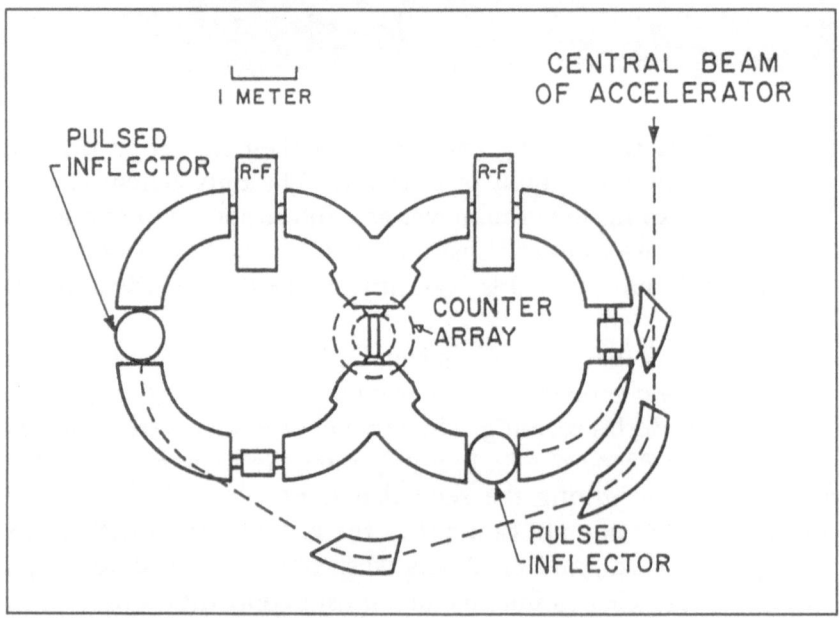

Fig. 8.2: The first electron-electron storage ring experiment. It was set-up by W. C. Barber, G. O'Neill, B. J. Gittelman, W. Panofsky and B. Richter at SLAC (Diagram from [O'N59]).

There was a place in Hamburg (the chamber of commerce [Am81]) where one could read foreign magazines and Touschek was a frequent visitor. This was noticed and the Gestapo arrested him in November or December 1944. He was jailed in Fuhlsbüttel, but was able to continue working for us there. We helped him as much as we could, but could not secure his release. I can remember that we brought his beloved books, some food and cigarettes to his cell, but I have no recollection of the schnapps he later spoke of. It was in prison that he wrote an important essay on 'radiation dumping in betatrons' which he wrote in invisible ink in the pages of Heitler's book 'The Quantum Theory of Radiation' [Am81].

As the British troops approached, Touschek was due to be transferred to Kiel in February or March 1945. He had a cold and was finding it difficult to carry his many books. One fell into a ditch and, while he was trying to pick it up (his condition was fairly poor), he was shot at from behind by one of the guards. He was only grazed behind the left ear, lost a lot of blood but was left for dead. When he heard passers-by speaking, he raised himself, was given treatment and then re-arrested and taken to Altona prison where, so he later said, things were 'a bit more peaceful'.

Touschek was freed by the British troops in June 1945. As I mentioned earlier, he then went to Kellinghusen where he wrote several interesting (theoretical) reports on the betatron [To45]. In several of them he developed ideas which I had suggested in our discussions, and had even submitted for patenting. His particular skill for theory and mathematical formulations was of great help. It was a very pleasant collaboration.

Touschek didn't publish this work and never mentioned it in his curriculum vitae. However, these reports must have come in useful when he went to Göttingen in early 1946. At about that time Konrad Gund's 6 MeV betatron built at Siemens Company in Erlangen was going to be installed at Göttingen. By the summer of 1946 Touschek had completed his thesis (it was on the theory of betatrons) under the supervision of Professors R. Becker and H. Kopfermann.

Box 10

Vacuum in Storage Rings

Gaede's vacuum pumps of Wideröe's Karlsruhe time could at best maintain a residual gas pressure of 10^{-6} millibar in a well sealed container. This is just about enough to run betatrons, cyclotrons, smaller synchrotrons and linacs. The acceleration process in each of these machines is completed within a fraction of a second.

The situation changes completely when particles are to be stored in a ring for longer periods. This necessitates an improvement of the vacuum of at least a factor 100. It quickly became apparent that such machines could only be built in places where particularly good vacuum experts were available.

The situation changes again when 10^{10} high energy electrons or positrons are to be stored, as is the case in some rings. These particles create so much electromagnetic ('synchrotron') radiation, that the temperature of the vacuum chamber increases dramatically. Therefore the gases which are enclosed at the surface, escape. The heat may be drawn off by water cooling, but the gases have to be pumped out. It may take several weeks before the vacuum is sufficient to run a storage ring, that is, before the vacuum attains 10^{-8} millibar when the beam is circulating.

Synchrotron radiation of protons is negligible (at the energies which can be reached today), and causes practically no rise in the temperature of the vacuum chamber and therefore no vacuum problems. In the HERA storage ring at DESY [Wa91] most of the 6.3 km long vacuum chamber for the proton beam is even kept at less than -268^0C. This has a similar effect to the so-called 'cryopumps': any remaining gases condense on the surface. The vacuum becomes so good that it can no longer be measured. This corresponds to about 10^{-11} millibar or better. The mean life of the proton beam then reaches several hundred hours.

Many technical and industrial innovations were necessary to achieve such progress in vacuum technology and thus to make possible the construction of modern storage rings. Almost all the parts used nowadays for these constructions are made of metal. Plastics, oil and mercury all belong to the past. Vacuum-tight welded and braced joints and flanges are generally used. The search for leaks in so called 'ultra high vacuum' systems is today a highly skilled profession.

Afterwards Touschek went to Glasgow where he obtained a PhD in November 1949. From December 1952 Touschek worked at Rome University. As a theoretical physicist, he made important contributions and wrote many very interesting publications during the course of his life.

But Touschek was also the first to break the ice in the field of storage rings. In Rome at the beginning of 1960, he proposed the construction of an electron-antielectron storage ring [To60]. This was completed within less than a year at the Laboratori Nazionali di Frascati in the beautiful hills to the South of Rome. It was the first storage ring ever to function, so it was the first time my patented ideas of 1943 were actually used in practice.

Electrons and their antiparticles (the positrons), have exactly the same mass, but electrical charges of opposite sign. They can therefore run in the same ring (and magnetic field) on identical orbits but in opposite directions and will then meet in certain places. According to Touschek (and to my patent), these encounters could eventually result in frontal collisions. Touschek's rather theoretical ideas were put into practice in Rome by brilliant experimental physicists.

In two other projects, similarly small storage rings of different types, were also built. One was in the USA, prompted by Gerry O'Neill [O'N56] (see Fig. 8.2) and another in Akademgorodok near Novosibirsk (then USSR). Construction of these two had started before, but they did not become operational until after the Frascati storage ring. In each of these two cases, two electron rings were placed tangentially next to each other. An interesting experiment was conducted on the American rings to check the validity of quantum electrodynamics.

After my 1943 patent, I was never really involved in storage ring construction, instead I concentrated on betatrons, a realistic task by then. But I did meet Touschek several more times, the last time was in 1975. He died of liver failure in 1978. He had been rather too partial to a drop of alcohol, and that was probably his undoing.

Touschek's machine in Frascati was primitive, but also interesting. It was given the name Anello d'Accumulazione (AdA) which in Italian corresponds precisely to the term 'storage ring'. As I mentioned earlier, a single ring was used to store both electrons and positrons in opposite directions and to make them collide. Basically it was two storage rings within a single tube, exactly as I had proposed in my patent of 1943.

However, a storage ring is merely a synchrotron with particularly good stability. I shall return to this subject later on. In AdA the particles could be stored at approximately 200 MeV. The machine as a whole had an external diameter of only 1.6 m, and the electron orbit was about 4 m in circumference. AdA went into operation on February 27, 1961. Touschek would spend hours watching a few stored electrons through a small telescope. A single electron gives off so much light during its orbit that it becomes clearly 'visible' (this is part of the so-called 'synchrotron-radiation'). AdA was later taken to Orsay, south of Paris, where positrons were also injected into it and made to collide with electrons [Am81]. The performance of AdA is best described in the PhD thesis of the Orsay physicist Jacques Haissinski [Ha65].

The development of storage rings led to gigantic machines which were often close to the limits of the available technology and financial resources. They were used to make very important discoveries, especially in relation to the quark structure of matter. The 'Large Hadron Collider' (LHC) should make collisions between protons of 8 TeV (1 TeV = 1,000 GeV) feasible at CERN, in the 27 km long tunnel of the electron-positron collider LEP. Two proton storage rings will be fitted into the same tunnel and the beams will be guided towards each other at various points. This machine should help to solve some of the important remaining problems regarding the structure of matter.

Let's get back to my life story. By 1945 I had problems of quite a different nature to contend with, especially after I returned to Norway.

9 Oslo – the Theory of the Synchrotron

After the German troops left Norway in May 1945 and the Crown Prince returned, I was arrested in Oslo and taken to Ilebu prison. The Germans had previously used their buildings as a concentration camp which was known as 'Grini', and many Norwegians still have sad memories of this place.

Sometime later I found out that one of my neighbours had reported me to the police because he knew about my expertise in regard to relays, and therefore believed that I had participated in the building of V2s in Peenemünde, and may even have invented them. Of course, that would have been a grave matter. We must not forget that London and Antwerp were still being attacked by V2s in April 1945 and there was no means of defence against them. During the winter of 1944/45 a total of 2,800 such rockets were launched over these two cities, each one carrying a ton of explosives – however, only a fraction reached their destination!

Luckily I had brought with me all the papers and documents related to the betatron construction in Hamburg. These enabled me to write an extensive report while in prison. And, when I completed it at the beginning of July 1945, I was released. Apparently I was helped by the famous accelerator expert Odd Dahl, whom I didn't know at the time. He had influential connections, but I imagine that a few others also had a hand in my release.

Although, as I mentioned earlier, NEBB (a member of the Swiss Brown Boveri Group) in Oslo were my employers for the entire duration of the War, and I had been 'conscripted' to work in Germany, I do not believe that this fact contributed to shorten my stay in prison.

From prison I wrote a long letter to my wife, making plans for the future. NEBB had stopped paying my salary after my arrest, and I was worried about my family. I asked my wife to pay a visit to the director of NEBB in Oslo and to ask him for advice. He

Box 11

The Experts' Report

Professor Roald Tangen writes from Oslo about the circumstances of the years 1945 and 1946 [Ta93]:

"I applied (in 1993) to the Norwegian National Archives for access to the documentation of that time. There I found an extensive file on Wideröe (several hundred pages – I could not go through all of them) including a copy of the experts' report which was compiled at the time, a document of about 15 pages from which I had a microfilm made.

The matter had first been handled by police officers in a committee of enquiry. From the documents it is apparent that the policemen had very little knowledge of nuclear physics and nuclear weapons and, accordingly, were not in a position to know whether a betatron could be used as a weapon of war.

Because of this, in November 1945, the police officer in charge (who, incidentally, was positively disposed toward Wideröe) called for a commission to act as advisors to the authorities regarding technical matters. The members of the commission were Professor Egil A. Hylleraas, Professor Harald Wergeland, Gunnar Randers and myself. Apart from myself, all have since died. Professor Hylleraas wrote the final text of the report [Hy46].

The papers in the Archives document that the work of the commission effected that the first charge, which concerned Wideröe's involvement in the construction of the V-bombs, was declared groundless. This meant that the charge was reduced to the general one of having worked for the forces of occupation.

The 'commission of experts' had no role at all during the legal procedure which took place much later (in November 1946).

I also found the concluding document of the case in the files, a 'forelegg', a kind of 'submission of evidence' for minor offences. (Wideröe accepted this 'forelegg', and in compliance with Norwegian law, his acceptance meant that the case was closed without a formal court trial [Wa94].)

After his release on July 9, 1945 Wideröe was not issued a passport at first, but later (in the spring of 1946) he was given a one-month passport so that he could go to Switzerland to join in the preparations for the construction of betatrons for hospitals."

suggested that I should apply for a position at BBC in Baden, Switzerland.

During my imprisonment I was also visited by the Norwegian physicist Gunnar Randers. He had spent some time in America and returned to Norway to devote himself to astro and nuclear physics. He was sent to talk to me, presumably because of the V2 rumours. It would be easy to check the date because it was the day of a solar eclipse and he had brought with him some blackened glass with which we could observe the sun. It was on July 9, 1945, about an hour after midday. I was given the opportunity to explain the facts about my activities in Germany and, at least in my opinion, he and I got on rather well.

A while later, a commission of experts was called to make a 'professional assessment' of my activities and to clarify my position [Hy46]. I didn't notice much of these investigations myself but I am not very susceptible when it comes to this sort of thing. It is possible that some people made malicious statements about me, but either I did not understand them or I wasn't bothered.

I assume that the police authorities just wanted experts to answer questions they were not in a position to assess themselves. I think that's quite natural under the circumstances, but the mood in Norway was a little overheated at the time and things were not always thought through and considered calmly and justly. In any case, it does appear that there were serious doubts about my conduct during the War. I bear no resentment, but at the time I did appreciate that I would soon be leaving for Switzerland to continue my work.

Despite everything, the post-War suspicions did leave a certain after-taste for some people and I am glad now that everything appears to have been completely cleared up. And the flowers I received from the Royal Norwegian Ambassadors during the last years on the occasion of various honours have entirely convinced me that no one in Norway now thinks badly of me. I was always very proud of being Norwegian. I was frequently, and mistakenly, described as being German, the first time probably in an article by

Professor Gustav Ising, which appeared in the Annuary of the Swedish Physics Association [Is33]. This must have caused the confusion.

My wife has a very clear memory of the second half of 1945 and especially of the winter of 1945/46. We had very little money, it was extremely cold, I had no passport and was practically unemployed in Oslo. I used this time to order and write down my thoughts on what was later called the 'synchrotron'. I submitted these ideas and theories as a private patent in Norway on January 31, 1946 through the agency 'Tandbergs Patent Kontor' Oslo [Wi46]. I had heard some rumours about other people in the USA working on the same subject, so I did my best to finish as soon as possible. The text of this patent is rather complicated and contains many formulae which I can no longer understand today. But it also includes some very important ideas which I shall describe later.

A synchrotron is made of a ring-shaped vacuum chamber, on which a magnetic field is applied. This field increases with the energy of the particles and keeps them on the same orbit. In such a machine, part of the beam-pipe corresponds to a bent drift-tube, as illustrated in fig. 5 of my patent (see Appendix 2). At the ends of this tube the particles receive voltage kicks every time they pass through, that is, once at each revolution. During this process the particles are automatically kept together in small bunches, which practically 'ride' on the accelerating wave. The frequency of the accelerating voltage and the speed at which the particles turn around must obviously match each other exactly. In a stable bunch the particles 'oscillate' around their nominal position inside the bunch, as they are constantly being 'pushed back' by the accelerating wave. These are the so-called 'synchrotron oscillations'.

The history of the invention of the synchrotron is very interesting. The idea must have been floating in the air at the time. In America, Edwin M. McMillan discovered the most important principle which was published in a very elegant article in the September issue of the 'Physical Review' 1945. It was merely two pages long and became world-famous [Mc45].

At almost the same time in Moscow, Vladimir Veksler quite independently discovered the same principle [Ve45], and he described it in an extensive essay. Oliphant and his colleagues in England also appear to have found the principle (at least, part of it), again on a completely independent basis. And when I submitted the above mentioned patent I had no concrete knowledge of the others. I saw McMillan's publication only a few months later. Scientific contact and exchanges of information were much disrupted during the War.

My patent was based on the drift-tube which I had extended to a ring shaped machine which I called 'λ/2' and 'λ/4' resonant accelerators. But the patent also contained many other very important details which are now taken for granted when a synchrotron is built. For example, the stipulation that the accelerating frequency must be exactly fixed by the revolution frequency of the particles (a very important condition) is included therein.

Much later, during the construction of the 30 GeV proton synchrotron (PS) at CERN in Geneva, one of the designers of this machine, Dr. Christoph Schmelzer (whom I knew well), proposed a different solution. He wanted to adjust the accelerating frequency to the revolution frequency of the protons using a computer. This, however, did not work. It was only after he had rigidly coupled the two frequencies that the machine was successful. Schmelzer called this a 'phase-lock' and it became one of the most important construction criteria for further synchrotrons.

The magnets of a synchrotron only produce an adequate magnetic field in the relatively small ring shaped region of the particle-beam and not in the central part of the orbit, as had been necessary with both betatron and cyclotron. The significance of this is that a synchrotron machine for a given energy costs less – or that with the same amount of money it is possible to build a machine for much higher energy.

Another interesting idea was the proposal for using a multiple of the revolution frequency in the acceleration. In doing so, the amplitudes of the synchrotron oscillations could be made smaller.

This was very important for further reducing the size of the vacuum chamber and also of the bending magnets.

In his famous article McMillan announced that his institute, the Radiation Laboratory at the University of California, was already planning to construct a synchrotron for 300 MeV, and this actually went into operation in January 1949. However, back in 1946 in England, the two physicists F. K. Goward and D. E. Barnes [Go46] had already undertaken a precise verification of the synchrotron principle with a modified betatron. They constructed an acceleration drift-tube made of wire mesh around the beam-pipe (complying exactly with the conditions specified by McMillan as well as in my patent) with which they were able to continue accelerating the electrons after the betatron-action was finished. They achieved 8 MeV, which was twice the energy previously reached with the same machine as a betatron. The way towards further developments was already apparent at this very early stage!

For the sake of completeness, I should like to mention the principle of 'strong focusing' which, although developed later, today belongs to the foundations of modern synchrotron construction.

It was at the beginning of August 1952 when, on my return voyage from Australia (where I had been lecturing on the betatron) I was travelling through the United States and got to Brookhaven. Here I met Odd Dahl and Frank Goward who had been sent over from Europe by CERN. We spent several days with Ernest Courant, Hartland Snyder, Stan Livingston and other interesting people who had developed the so called 'strong focusing' method just a few weeks previously [Co52]. With the help of magnets of different form ('alternating gradients'), this allowed for further reductions in the beam's dimensions and therefore also in the size of the vacuum chamber. Thus beam focusing was strengthened, that is, the particles were better bundled. Following this principle it became possible to build larger accelerators at the same cost.

Whilst in Hamburg, I had proposed another method for improving the focusing of particle beams, the 'lens-road', which I submitted as a patent in September 1943, just as we were starting to build our betatron. I had been thinking about this problem for quite a while. But in the end the 'alternating gradients' were much simpler to realise and I would say that they were also more effective – they were just better.

Incidentally, the first inventor of this method was the Greek Nicholas Christofilos who had had this idea patented in March 1950 [Ch50]. However, it was not accepted and published until February 28, 1956. He worked for Westinghouse and I met him once during a conference in Russia.

Now I would like to go back to my life story. In Oslo I was given a passport just before Easter 1946 and I flew to Switzerland for a short visit where I met the very pleasant Professor Paul Scherrer. Years later, the large Swiss research centre in Villigen was named the 'Paul Scherrer Institut' (PSI) after him. It is quite close to where I live now in Nussbaumen and not far from Baden (CH).

During this trip I also met one of the Boveris. I think it was Walter Boveri, although it may have been Theodor Boveri. We agreed that I would build a fairly large betatron for Brown Boveri (BBC) in Baden. In those days we did not believe that this could be done in Norway. The infra-structure was not suitable; for example, there were no adequate glass-blowers and no vacuum technology to speak of.

10 Baden – Betatrons for BBC

In the spring of 1946, after my position in Norway had been sufficiently cleared up, I went to Switzerland and started on some preliminary work for a betatron. The construction drawings produced at that time were already pretty detailed. This was a machine with which electrons were going to reach energies of up to 31 MeV, which corresponds to an acceleration by 31 million volts. We had opted for 31 MeV because we wanted to extract the electrons out of the vacuum chamber, as we later in fact did. Electrons of 31 MeV penetrate about 10 cm of water – or the equivalent body tissue – and could therefore eventually exert the appropriate therapeutic effect. The machine was conceived first and foremost for such medical purposes.

The iron yoke consisted of six return sections arranged in the shape of a star, as is shown in Fig. 10.1. This was a construction which I knew well from the manufacture of transformers. The six sections were made of iron plates which had been soldered together. Mr. Hartmann and a few other members of the BBC's staff then went on to build the machine according to my instructions.

My wife, our three children and I left Oslo for good on August 19, 1946. We went with our car, first taking a ship to Amsterdam and then driving to Zurich via Luxembourg. In Zurich everything seemed to happen very casually. I somehow obtained a work permit – I don't know how myself. Apparently it was a case of 'established facts' which had been taken care of by BBC.

As Ragnhild remembers very well, I had to return to Oslo in October to clear up the case about my work in Germany during the War. I stayed with my parents while I was in Oslo. Because I accepted the confiscation of the last money earned in Germany, no trial was required and I was subsequently given another passport. In November I was permitted to return to Zurich.

The betatron started to take shape at the beginning of 1947, and the manager of BBC's department in Baden allocated us a 'working place'. It was a section of a tunnel beneath a large hall used for testing generators. It was one of the tunnels through which the warmed cooling air and innumerable other vapours were extracted and was also used to inspect the big machines from below.

Such was the tunnel in which we were supposed to construct the betatron. Working conditions were very bad. Above our heads were the big machines and, when they were running, it was impossible to hear anyone speak; we would be forced to flee. Now and again the generators' coils were impregnated with various insulating substances and then it would become impossible to breath as the vapours were extracted through our tunnel. However, we managed to do some work regardless.

But we had difficulties getting the betatron to work. This was because the yoke's six iron return sections were not exactly identical. This meant that the magnetic fluxes for each section went through zero at slightly different times and provided different steering fields for small currents. And this happened very close to the moment in which the electrons would have to be injected. A successful injection of electrons was thus very rare indeed.

The Hamburg machine only had two returns and was therefore easier to adjust. Kerst's second machine (in the USA) for 20 MeV also only had two return sections. However, it didn't take us terribly long to find a solution (it was sometime around January 1948). On each of the six yoke returns we fitted ten copper windings which were short circuited via an adjustable resistance. By doing this we were able to optimise precisely the steering fields at the moment of injection. The fields were accurately measured by means of small permalloy strips fitted above the air gap. Incidentally, the six yoke returns proved to be rather a boon, because they screened off a large proportion of the high energy X-rays produced when the machine was running.

The hazardous spatial conditions soon made us subject to high doses of radiation as we didn't have enough space for shielding.

103

We therefore had to drive to the Kantonsspital in Zurich once a week to have our white blood cell levels checked. If we had less than 3,000 per cubic millimetre we would have to take some time off. After we increased the power of the machine even further, the radiation became too high even for the workers on the level above. This factor effected an important improvement: we were finally provided with a proper 'radiation laboratory' in which we were able to protect ourselves from the radiation.

BBC gave me a free hand and practically all decisions were left to me – except, of course, with regard to our place of work. This was because I was the only one who had any understanding of betatrons. Initially I was just told to build a betatron, and this was mainly thanks to Professor Scherrer who had been an ardent campaigner for the construction of such machines. His interest was probably decisive. Furthermore, BBC wanted to be 'on the scene' of nuclear and particle physics; the betatron was going in that direction although, at the beginning, its only purpose was medical. Perhaps those 31 million volts had a certain hypnotic effect. And the atomic bombs which had exploded over Japan had raised the industry's awareness of nuclear physics.

I would like to mention again the support we had from Walter Boveri. He was a good friend of Professor Scherrer. Later, although not very much later, Dr. Hans Rudolf Schinz of the University of Zurich also entered the scene and he turned out to be a great advocate for the construction of betatrons. He ran the Radiotherapy Department at the Kantonsspital in Zurich.

Apart from their medical uses, the betatrons also became important for the non-destructive testing of materials. Even the 15 MeV betatron from Hamburg was used for this purpose after it was shipped to England.

When we had made ourselves comfortable in the new radiation laboratory, we progressed quickly, and in autumn 1949 we took the machine to the Kantonsspital in Zurich where a specially equipped room was ready and waiting for it. There was still much to do, especially with regard to the radiation shielding. Many

104

Fig. 10.1: Diagram of a BBC betatron with its six return yoke sections, as shown in one of Rolf Wideröe's patents [Wi49].

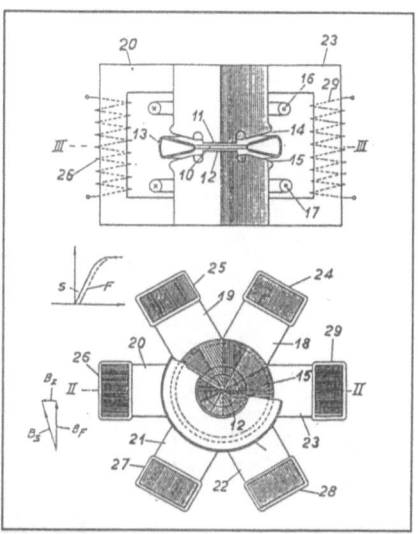

Fig. 10.2: A BBC 31-MeV-betatron during construction. Mr. Gamper (left) and Rolf Wideröe (right). (Photograph BBC).

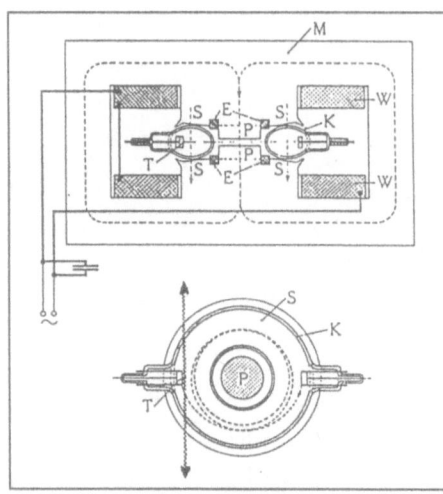

Fig. 10.3: Diagram of BBC's double-beam betatron.
M = Magnet yoke
P = Central magnet poles
S = Steering poles
W = Exciting coils
E = Expansion coils
K = Ring tube
T = Anticathode (target)
[Wi62].

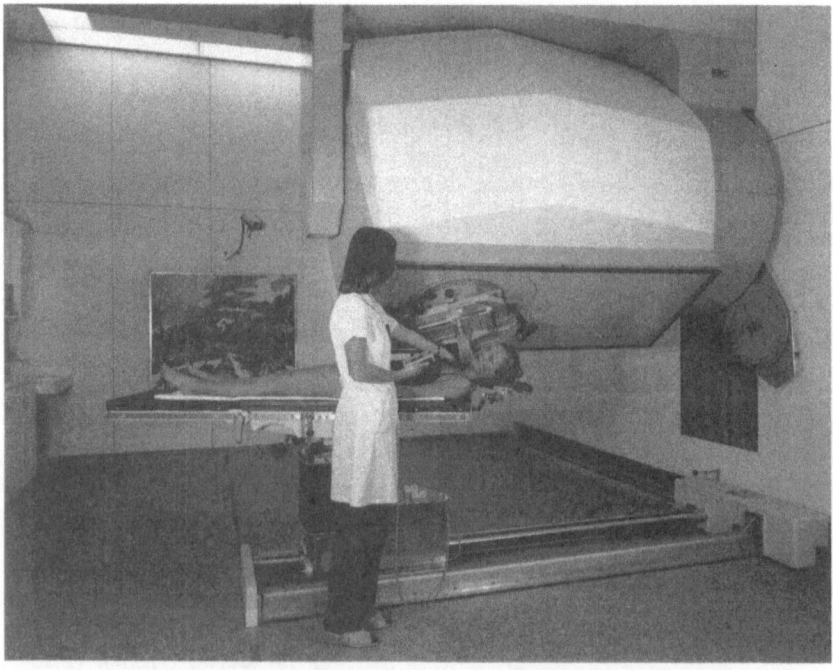

Fig. 10.4: Betatron radiation therapy, Inselspital Berne (phot.: BBC).

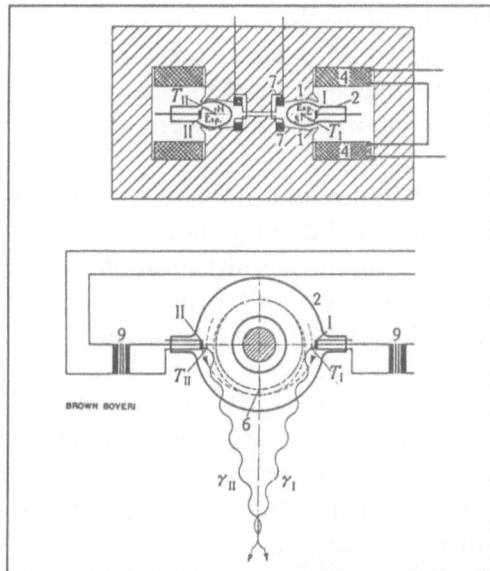

Fig. 10.5: BBC stereo two-beam betatron for materials testing.
1 = Magnet pole
2 = Ring tube
4 = Coil
6 = Orbit
7 = Expansion coil
9 = Impulse transformer
I+II = Electron sources
$T_I + T_{II}$ = Targets
$Y_I + Y_{II}$ = X-rays
[Se58].

Fig. 10.6: A betatron being used to test a Pelton-wheel at Georg Fischer AG, Schaffhausen (photograph: BBC).

107

measurements were taken and a lot of shielding had to be fitted to protect against unwelcome X-rays and even against neutrons, which this type of machine also produced. Lead plates served as shields and later on substances containing boron were also used for these purposes.

I remember one day Professor Schinz came along to see us with a visitor. At the time I happened to be lying underneath the machine. He pointed his walking stick in my direction and said, "there lies my greatest enemy". We weren't progressing fast enough for him. But we did complete the machine eventually, and the first patients were given X-ray treatment in April 1951.

By 1952 we were in a position to deliver a further two betatrons, one to the Inselspital in Berne and one to the Radiumspital in Oslo. With regard to the latter hospital, I would like to say a few words more. My friend Olav Netteland told me that Dr. Johan Baarli (who later became the head of the Norwegian Service for protection against radiation) measured the number of neutrons in the surroundings of the machine and found that it was far too high. He even called it 'Wideröe's sterilisation machine' – if I remember rightly. However, Baarli had not taken into account the difference between fast neutrons, which are dangerous, and slow neutrons, which are relatively harmless. Nevertheless, someone at the hospital had claimed that he was suffering from headaches... It is my opinion that most of the measurements made at that time were plain and simply wrong.

We did have some protective regulations regarding radiation, but they had not yet been defined very precisely. The permitted radiation doses were about five times today's top limits, which are really quite low. The people of Kerala in southern India live under constant exposure to radiation doses which are five times higher than those permitted by our regulations. The cause is monazite sand containing radioactive thorium. Nevertheless, the local population does not appear to have suffered adverse effects.

As I can remember very well, that the instruction sent to us by Oslo's Radiumspital was the most unusual Brown Boveri ever received. The head of the hospital, Dr. Reidar Bjarne Eker, simply wrote us a letter with the words, "We order a betatron", his signature and the date. Not a word about energy specifications or any other data. We built him a 31 MeV X-ray betatron.

In 1956 we managed to extract the electrons from the glass tube of our betatron. We did this by using a process for which I had submitted a patent several years previously [Wi52]. In 1957 we converted a betatron, which we had delivered to the Inselspital in Berne in 1953, so that it would function with this supplement. Special coils were fitted in the air gap above the ring tube. They were called 'pancake coils' because they were so flat.

By the way, most of our betatrons were able to deliver two beams simultaneously, emitted in opposite directions, as shown in fig. 10.3. The tube was thus exploited more efficiently as particles were accelerated during both the positive and the negative rise of the alternating current. This made it possible to treat two patients simultaneously in separate rooms. Naturally we had to provide the machine with electrons which moved in opposite directions at injection, as well as ensuring that the electrodes on which the X-ray beams were produced were shaped appropriately

An interesting variation was developed for the non-destructive investigation of large industrial components. The X-rays were produced at two opposite points in the tube so that the target object could be X-rayed simultaneously from two different directions, thus making two stereo pictures of the interior (see fig. 10.5). We were able to reduce the size of the 'sources' of the two X-ray beams to a few tenths of a millimetre in order to achieve a better photographic resolution. We had got the hang of it and our betatrons were probably among the best industry could produce.

It may be interesting at this point to mention the development of radiation therapy at the Radiumspital in Oslo, because similar processes were also taking place in other countries. Initially, a generator for high voltages was due to be built in Bergen during the War, a 'Van-de-Graaff machine'. Odd Dahl describes all this very nicely in his book published in 1981 [Da81].

Box 12

Betatrons and Industry

Following Kerst and Serber's first publications in 1941, industry had a good sense of the potential demand for betatrons, both for medical purposes and for non-destructive materials testing (less so for basic research in nuclear physics). Even during the War interesting developments were being initiated in both Europe and America, mainly in view of the commercial market expected after the War.

The American companies General Electric (where Kerst built his 20 MeV betatron in 1942), Westinghouse (from whence Slepian submitted the first patent for a preliminary stage of the betatron in 1922) and Allis-Chalmers devoted their attentions to the commercial manufacture of 20 MeV betatrons.

In Europe, Konrad Gund developed and built 6 and 15 MeV betatron-machines (which later achieved 18 MeV) at the Siemens-Reiniger factory in Erlangen, following Max Steenbeck's ideas and suggestions

In 1946, Wideröe started to develop and produce the 31 to 45 MeV machines which were such a success for Brown Boveri & Co. (BBC) in Baden, Switzerland.

Philips' interest in the betatrons had already become apparent in 1944, when Wideröe worked with the company C. H. F. Müller in Hamburg (which formed part of the Philips group). Later on, betatrons both with and without iron cores were also built by Philips in Eindhoven. The iron-less betatrons for 9 MeV were run in a pulsed mode. Philips and BBC seemed to maintain good relations as was demonstrated during the production of electron sources for Wideröe.

In an article written in 1962 [Wi62] Wideröe described the three types of betatrons for hospitals which had been developed and built by Siemens-Reiniger in Germany, Allis-Chalmers in the USA and BBC in Switzerland. He also described the interesting linear accelerators developed at that time which could be used for medical purposes.

It is difficult to estimate the precise total number of betatrons built throughout the world. Commercial firms probably installed more than 200 of them, of which 78 were manufactured by BBC. But many institutes developed and built their own machines.

First they tried to get the machine built by Philips in Holland, but this proved too expensive; they had managed to collect just about 150,000 Kroner, and that wasn't enough. We must remember that such a 'Van-de-Graaff' could replace the radiation of a kilogram of radium. And in those days, as I already mentioned, one gram of radium cost around one million Kroner!

Philips had recommended that they should build the machine themselves, especially since Odd Dahl could be in charge of the technical process. He had already successfully built and operated high voltage machines in the USA.

The Van-de-Graaff machine was completed in 1941 in an extension of the Bergen Hospital. It reached 1.7 million volts. After this, Dahl supervised the construction of another machine (of the same type) at the hospital in Haukeland. This one even reached two million volts. Finally, the Radiumspital in Oslo asked for a similar machine and construction began.

However, when betatrons became available on the market in 1948, the head physician at the Radiumspital, Dr. Bull-Engelstad, ordered one from Siemens in Erlangen. It was to have an energy of 6 MeV and was scheduled for delivery in 1949. As already mentioned, Siemens had been developing this type of machine since 1941. The parts of the Van-de-Graaff machine under construction were given to the University of Bergen as a gift.

This was the situation as Olav Netteland found it when he began working at the Radiumspital in September 1949. In the autumn of the same year, Netteland went to Erlangen to take a look at the betatron. By then Siemens was already developing a 12 or perhaps even an 18 MeV betatron.

At that time we at BBC in Baden had progressed quite far with the 31 MeV machine for the Zurich hospital. A congress of radiologists took place in London in 1950 where Siemens exhibited their 6 MeV machine. However, it emerged later that this was a non-functional exhibit and hadn't even been fitted with a tube. After this, Olav Netteland contacted me, and in September 1951 he and head physician Dr. Steen came to Switzerland to see our

111

31 MeV betatron which was already up and running at the Kantonsspital. I went to Erlangen in the autumn of the same year where Siemens could only show me the 6 MeV betatron. Completion of the 12 MeV machine was still a long way off. I did not have any difficulty in having the Siemens order cancelled and Prof. Eker immediately ordered 'a betatron' from BBC. We delivered a 31 MeV machine in the summer of 1952. Within six months it was operational. I think Siemens did provide the Radiumspital with a machine eventually, but I don't know much about that.

After that I was in frequent contact with the Radiumspital in Oslo, especially with Professor Eker. I have kept my letters of that period. The hospital did not provide radiation therapy until 1953, and in the first few years we had a few problems. The cathode of the electron source had a very short life and we frequently had to replace the tubes. The 'oxide-cathodes' available at that time only run for about 500 to perhaps 1,000 hours. That was far too little. We experimented with other cathodes but our trials were not very successful. The barium aluminate contained in the cathodes attacked and dissolved the filament. Although Olav Netteland said that things were much better during the second year, the problem was not solved until I went to visit Philips in Eindhoven who suggested their own patented method. This was in the autumn of 1957.

After that, Philips supplied us with cathodes in the form of small tubes made of sintered tungsten powder impregnated with barium aluminate (which corresponded to approximately 30% in volume). We fitted these tubes with narrow cylinders made of aluminium oxide, each of which included a filament. We had to take great care to ensure that the filaments were completely protected by the aluminium oxide and that they could not come in contact with the barium aluminate, otherwise they would corrode and break very quickly. The most favourable temperature for the filaments was a little below 1,100°C. At this temperature, approximately as much barium oxide was diffused to the surface of the cathode as would be used up by ion bombardment.

These cathodes were very robust. Discharges did not destroy them, they regenerated themselves very quickly and they had an unbelievably long life, certainly well over 20,000 hours, perhaps even as much as 40,000 hours. We subsequently built betatrons which could run for more than 25 years without needing a new tube. Some of them are probably still in use today.

We had an excellent mechanic at BBC, Mr. W. Gräf, who knew how to execute the very precise work involved in building these cathodes. It is greatly thanks to him that our machines lasted so long. He also looked after the manufacture of the glass tubes. We had some very good people in our department who would help during the installation of the machine on-site. They would get it started and also assist in running it. We also undertook all repairs and supplied spare parts. From 1954 onwards I was in charge of 'EA', the Electrical Accelerators Department, which was renamed 'EKB' (Electrical Components for Betatrons) after 1973.

I would have to draw up a very long list if I were to mention all the colleagues who contributed to our success over the many years. I apologize for not being able to do this. However, I would like to call to mind just a few names, for example Dr. A. von Arx, Dr. M. Sempert, Dr. H. Nabholz, Mr. K. E. Drangeid (a Norwegian who later joined IBM's Research Laboratory), Mr. Gamper (he worked on materials testing), Mr. von Dechend (design engineer), Mr. E. Jonitz (head of the workshop) as well as Messrs. Vikene, Fischer and Gerber who took care of assembling and commissioning the machines.

The betatrons continued to be manufactured until 1986, by which time BBC had delivered 78 of them. I had submitted 53 patents for BBC, most of them in Germany but also quite a few in Switzerland. My time at BBC in Baden was, therefore, a very productive period. Towards the end of my career there I had put in applications for over 200 patents in all. A copy of each one is kept in the Archives of the ETH Library in Zurich [Wi70].

In 1959, we supplied the prototype of a 'mobile' betatron to the private hospital 'Casa di Cura S. Ambroglio' in Milan. The

Box 13

BBC-Betatrons from 1949 to 1986

Country	31-MeV Industry Research	31-35 Medic. (fixed)	Magn.- Lenses	Asclepitrons: (mobile) 35 MeV	45 MeV
Austria	-	-	1	2	1
Belgium	1	-	-	3	-
Canada	-	-	-	2	1
China	-	-	1	1	-
Czechoslovakia	-	-	-	-	1
Denmark	-	-	2	3	-
Finland	-	-	1	2	1
France	2	1	2	5	-
Germany	2	-	2	2	2
Greece	-	-	-	-	1
Hong-Kong	-	-	-	1	-
Israel	-	-	1	1	-
Italy	2	1	-	2	1
Japan	1	-	-	-	-
Jugoslavia	-	1	-	-	-
Norway	-	1	-	1	1
Spain	-	-	-	-	2
Sweden	-	-	-	4	-
Switzerland	1	2	2	2	5
U.K.	1	-	1	2	-
USA	-	-	2	5	7
USSR	1	-	-	-	-
Totals:	11	6	15	38	23

In all, 78 installed betatrons and 15 magnetic lenses.

Director, Professor P. L. Cova, placed the order with us and the machine was still in use a few years ago. This machine 'revolved' around the patient. We christened it 'Asclepitron' after the Greek god of medicine 'Asclepius' or 'Aesculapius'. As of 1967 we were able to increase the energy of our betatrons to 35 MeV and, in 1970, we even went as high as 45 MeV, which was of significance for particular applications in materials testing.

I had also developed a revolving 'magnetic lens' which allowed one to direct the electrons which came from different directions on to the spot which required irradiation. This minimised damage to healthy tissue. The lens also became a great commercial success. Many hospitals which ordered betatrons asked for them to come fitted with the magnetic lens.

After 1970 the demand for betatrons declined. By then it had become possible to build linear accelerators which were smaller and lighter than our betatrons. But above all they were cheaper, and in the end this was decisive. Important contributions to these developments came from the Stanford Linear Accelerator Center SLAC, frequently in collaboration with the company 'Varian'. This company recently took over the entire department at BBC which I directed. And in the meantime BBC had been renamed 'Asea Brown Boveri' (ABB).

I think that the most important machine after the betatrons which the BBC department under my direction developed was a synchrotron for Turin University, although a better name for it may be 'beta-synchrotron'. I would like to describe this in a bit more detail.

11 Turin – the Beta-Synchrotron

At BBC, the construction of betatrons in which electrons reached 31 or even 45 MeV was a great success. However, there were good reasons for not using betatrons to achieve higher energies, as I knew well from past experience. At the end of the War (1944), the German Air Force had appointed BBC to make preliminary plans for a 200 MeV betatron in accordance with ideas I had developed previously. I now believe it most unlikely that these proposals would ever have resulted in a machine capable of functioning.

Donald Kerst had already been successful in building and operating his second betatron (20 MeV) for General Electric in 1942 [Ke42]. W. F. Westendorp and E. E. Charlton then went on to build a 100 MeV betatron for the same company, and this was

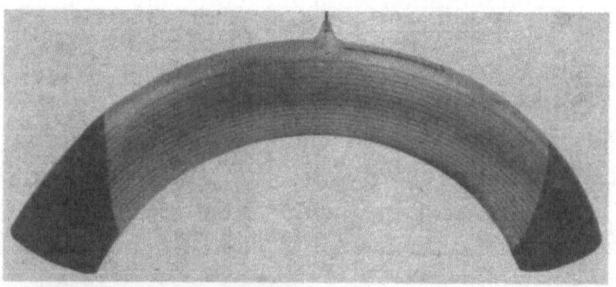

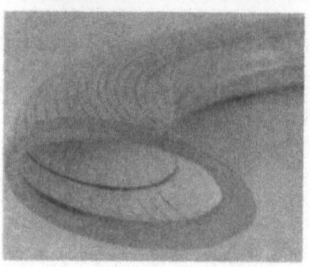

Fig. 11.1: The accelerating electrode of the Turin synchrotron [Go64].

completed in 1945 [We45]. In the meantime, Kerst had returned to the University of Illinois where he built first, a model machine for 80 MeV and eventually a gigantic betatron for 300 MeV. This was the largest machine of this type ever constructed and should be regarded as the final stage in the development of betatrons.

Size and construction costs prohibited competition at higher energies with synchrotrons which, by that time, had already been tried and tested. However, the betatrons proved their worth for practical uses below about 50 MeV for a long time. Later on, linear accelerators were developed for use at somewhat lower energies and these prevailed, especially for therapeutic uses.

After I started work at BBC (Baden) in 1946, we only ever discussed betatrons of 31 to 45 MeV. In the interim, I had also spent a lot of time thinking about other methods of acceleration, especially about that type of machine which McMillan called a 'synchrotron'.

From 1953 onwards I was several times in Italy to talk with various physicists about the construction of synchrotrons. Professor Giorgio Salvini and the engineer Fernando Amman were planning a 1,000 MeV electron-synchrotron at the time. This was later built in the 'Laboratori Nazionali di Frascati' south of Rome, where Bruno Touschek was also working at the time. This 1,000 MeV synchrotron came into operation in 1959.

Also in 1953 I entered into negotiations with scientists at Turin University's Institute of Physics for a new, much smaller, accelerator. My contacts at the Institute were the head, Professor Gleb Wataghin who came from Russia and had also worked in Brazil for a long time, and Professor L. Gonella. I have fond memories of them both. Incidentally, the project was financed in equal parts by the 'Consiglio Nazionale delle Ricerche' (CNR), FIAT in Turin and the University of Turin itself.

The purpose of this machine was to accelerate electrons to approximately 100 MeV, mainly for experiments in nuclear physics for which the secondary production of neutrons was also rather important.

117

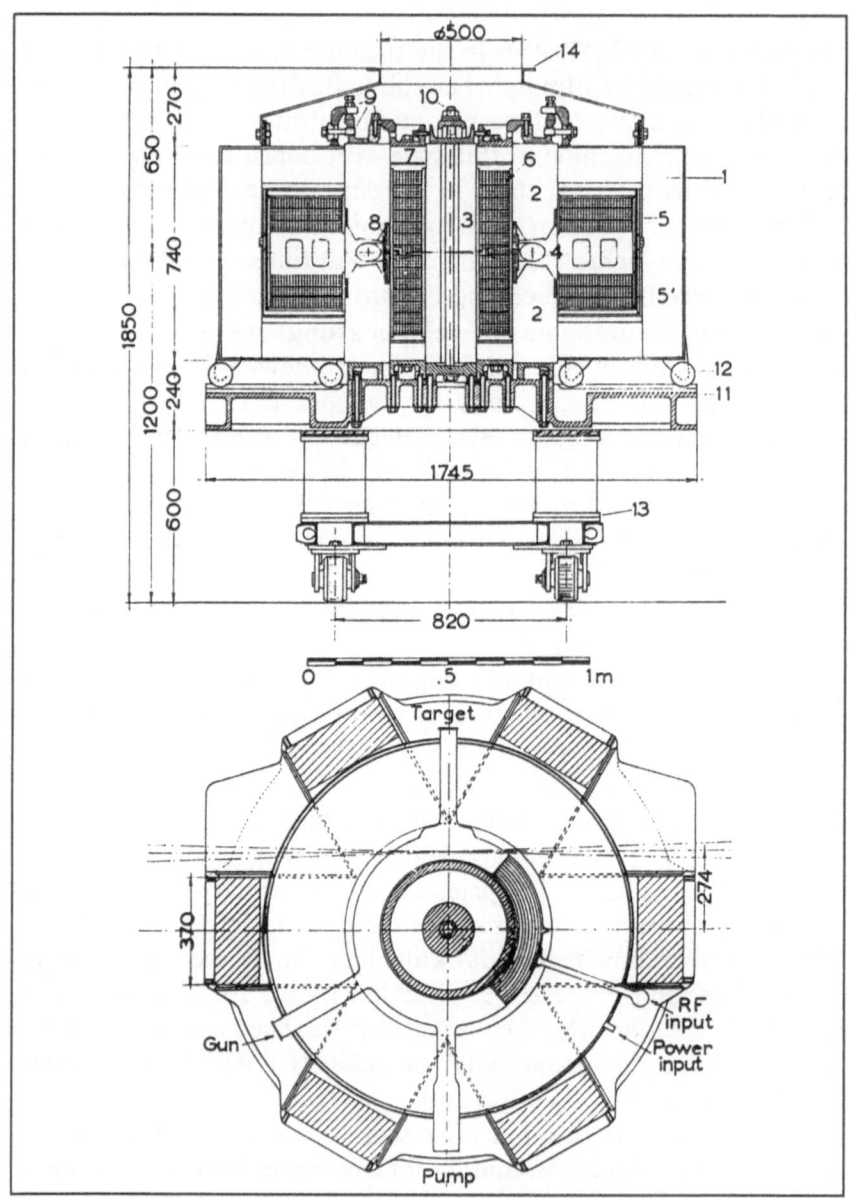

Fig. 11.2: Diagram of the Turin synchrotron [Go64].

118

It was clear to me by then that a betatron would not be the best solution for this task. Using the synchrotron principle we would be able to build a much smaller machine and achieve better results – at the target energy of 100 MeV. However, a synchrotron requires an injector, that is, a pre-accelerator which provides the particles with a starting energy.

The physicists at the Turin Institute were willing to tread relatively unknown paths in order to produce a very compact, reliable and economical machine which may also in the future be used at other places of research. So we developed a rather original concept, although it did owe much to the investigations done previously in the U.K. by F. K. Goward and D. E. Barnes [Go46]. Dr. H. Nabholz worked with me on both the design for the project and the construction of the new machine.

The machine was to function as a betatron until the electrons reached 2 MeV. Then it would continue to increase the particles' energy like a synchrotron. For me, this was the longed for opportunity to use my ideas and knowledge of synchrotrons on a machine which I was going to build myself.

And, of course, the new project was based on our previous, positive experiences of constructing betatrons at BBC. Accordingly, the iron yoke was again made up of six sections arranged around a central body. Naturally, many other details were taken from our betatrons, but the important thing for me was the second stage of acceleration, with which we hoped to achieve 100 MeV.

I had already described in detail the principles and the theory for the operation of a synchrotron in my Norwegian patent of January 1946 [Wi46] (reproduced in Appendix 2) and for the first phase of the operation (as a betatron), we were going to realise a few ideas which I had patented in 1948.

This machine was going to accelerate electrons in both directions, as was the case with many of our earlier betatrons. We fixed the radius of the electrons' orbit at 29 cm and planned to use the Italian electricity network's frequency, i.e. 50 cycles per second – which is what I had done with all my previous betatrons.

Part of the vacuum chamber was arranged as a curved drift-tube. In order to bring this about, a section of the inner surface of the chamber was coated with silver and connected to a high frequency voltage supply through a capacitor. As in my Aachen drift-tube, the electrons would be accelerated at both ends – but this time, the acceleration would occur once per revolution. However, this created many new problems and was not as easy as I had thought when I wrote my synchrotron patent in 1945 (see Fig. 5 in Appendix 2). This was not a simple drift-tube like the one I had tested in Aachen.

We had problems with secondary electrons which appeared on the inner wall of the tube. We dealt with this, and a few other problems, by coating the drift-tube with a layer of graphite (which has a high electrical resistance). We cut grooves along the coating and came up with a few other tricks, all of which we described in a subsequent publication [Go64].

By 1956 it became clear that we would need more time than originally expected to build the machine, so BBC provisionally installed a 31 MeV betatron in the Turin Institute. This was operated until the new beta-synchrotron was finally delivered.

When the 105 MeV machine was ready in 1959, the physicists of the Institute, and particularly Professor Gonella, were able to use it for many experiments. Gonella had also been active in installing and commissioning the machine. Together with my BBC colleague Nabholz, we subsequently wrote a report on the successful operation of the machine [Go64]. It contains many interesting details. More than anything it was important for us to demonstrate that such a machine had proven itself in practice, and furthermore, that it was relatively simple and cheap.

Even simpler and compacter linear accelerators were developed later for this range of energy and these have pushed aside both the betatrons and the small synchrotrons. Today these linacs dominate the market. However, developments are still possible and I assume that better and more compact machines will be built in future.

120

12   ETH Zurich, CERN and DESY

My inaugural lecture as an outside lecturer at the Swiss Federal Institute of Technology, Zurich (ETH) took place on December 12, 1953. It's subject was the history of particle accelerators. I prepared the lecture very carefully and I still have the original manuscript.

My lectures were always about 'particle accelerators'. Attendance was not compulsory and I had relatively few students. However, some were very studious and sharp. The most important aspect of these lectures for me was in the preparation. At last, I had peace to work through the theories of particle accelerators, and I collected a nice set of formulas which contained everything one would ever need to calculate particle orbits for the various types of accelerators.

I became 'Titular Professor' in 1963. I have good memories of my personal contacts at the ETH and I enjoyed my time there. I taught there until 1972.

But let's go back in time a bit, back to the year 1952 or perhaps even a little earlier. As far as I remember, the first definite proposals for a joint European laboratory for nuclear physics came from the French physicist Pierre Auger. From 1948 to 1959 he was the Director of the 'Department of Exact and Natural Sciences' at UNESCO which was then concerned with re-establishing research in Europe following the devastation of the War years. Some of the basic ideas may also have come from the famous physicist Isidor Rabi. During the second half of 1951 a 'council' (French, 'conseil') was set up for this purpose and many influential people from different European countries became involved. The name CERN is an abbreviation of the full title of the council, 'Conseil Européen pour la Recherche Nucléaire'. There is now available an in-depth report on these developments by A. Hermann, J. Krige,

U. Mersits and D. Pestre, the 'History of CERN' in two volumes [He87]. A very interesting 'Who's Who' section is included, which, however, contains a few mistakes regarding my life story (Vol. 1, p 565).

The idea of a research centre involving many European countries was discussed during a conference in Copenhagen in June 1952. This is where some projects were planned in greater detail. And I was there – simply because I was interested in this sort of thing. It had little to do with my work for BBC.

In Copenhagen we talked about constructing a synchrotron for 10 GeV protons. I remember some of the participants arguing that I was trying to push things too fast. I on the other hand felt that we should concentrate more on technical questions rather than get bogged down with administrative problems. However, I guess it was necessary to establish precise organisational principles before going any further.

The head of the planning group responsible for this type of accelerator (from which the proton-synchrotron 'PS' at CERN emerged) was Odd Dahl, who was working in Bergen and had a very good reputation as a builder of accelerators. He dedicated about a third of his time to the CERN projects. His deputy was Frank Goward. As I already mentioned, he and Barnes, were the first to successfully test the synchrotron principle. H. Alfven, W. Gentner and F. Regenstreif were also in the group, and they were later joined by D. W. Fry, K. Johnsen and Chr. Schmelzer. I was called in as a part-time advisor.

There was a second group, developing a 600 MeV proton-synchrocyclotron, CERN's future 'SC'. It was headed by the Dutch physicist Cornelius Bakker, but I didn't really have much to do with them.

After the Copenhagen conference I exchanged several letters with Kjell Johnsen to sort out technical questions on the planned proton-synchrotron. Most of the problems were new and their solutions still unknown. I sent Johnsen my calculations and reprints of some of my publications. I had made a mathematical error at one stage and Johnsen corrected this, but we got together a good basis for the construction of a 10 GeV machine. All my calculations were based on my Norwegian patent of 1946 [Wi46], that is, on my own synchrotron theory.

Soon after that I went to Australia and, as I mentioned earlier, I met in Brookhaven Odd Dahl, Frank Goward and the Americans who had invented the 'strong focusing' system on the way back. We spent a whole week in discussions, from August 4 to 10, 1952, and every bit of it was interesting. I understood immediately that their's was a much better idea than my previously proposed 'lens-road' for beam focusing. We decided thereupon to upgrade the proposal for a CERN machine to 30 GeV and to fit it with this modern 'strong focusing'. And, what's more, the Americans were prepared to help us in this somewhat risky pioneering work.

Looking back it is easy to evaluate the success of this adventure. At the Dubna Research Centre, in the North to Moscow, a 10 GeV proton-synchrotron was being built, still using the traditional method now called 'weak focusing'. Their machine was completed in 1957 and 36,000 tons of iron were used to build its magnets. This machine had the highest energy anywhere in the world at that time.

For our strong focusing machine at CERN, which eventually achieved 28 GeV, we only used 3,200 tons of iron, that is, less than a tenth of the iron used at Dubna.

This was definite progress – the risk had paid off. On top of this, CERN's machine went into operation before the one built by our friends in Brookhaven. It started operation on November 24, 1959 and it then snatched the world record for particle energy from our Soviet colleagues.

Naturally, we had various problems to deal with, starting with the planning stage. One difficulty consisted in making sure that the betatron oscillations of the protons did not come into resonance with the revolution-frequency of the protons. Because of the large oscillation amplitudes, this would lead to the loss of particles which would then hit the wall of the vacuum chamber.

While a consultant at CERN between 1952 and 1956, I was not given any particular guidelines or set tasks. The consultancy took up a small portion of my time, most of which I spent at BBC in Baden building betatrons. The latter occupation was barely connected to the great CERN project, except perhaps by the fact that a few electrical machines which were needed to excite the CERN magnets were later supplied by BBC. The kinetic energy stored in these machines was then discharged through the magnets.

I received copies of all documents, calculations and comments which were produced for the CERN machine. Every now and then a meeting would be called. For example, on December 18, 1952 I went to Geneva and together with Professor Gentner and Dr. Citron, I visited the site where the machine was going to be built. Citron was working on the planned accelerator's screening. He later became a professor in Karlsruhe. On that occasion, we did choose the sense in which protons had to turn in the machine, in such a way that farms or villages would not be affected by any particles which may escape tangentially. A protective hill had to be thrown up later and it became known as 'Mont Citron'.

I remember working on many interesting and useful calculations with Frank Goward, Hildred Blewett and other members of CERN's staff. Now and again I would meet Odd Dahl and Hildred Blewett – they were good friends. When Odd Dahl returned to Norway for good, John Bertram Adams (from England) came over to Geneva and Kjell Johnsen became his right hand man. But there had been others at CERN before Adams, such as Cornelius Bakker, Lew Kowarski and also Viktor Weißkopf. I knew Weißkopf pretty well.

Odd Dahl has written a very nice book which I mentioned earlier. He describes many things about the beginnings of CERN in this book which is written in Norwegian [Da81] and is now regrettably out of print. On p. 191 he writes that one of his friends helped me get to Switzerland after the War. I think he may be referring to Gunnar Randers. As one of the Norwegian delegates, Gunnar Randers was also very active for CERN.

Fig. 12.1: View inside the CERN-PS tunnel (Photograph: CERN).

Fig. 12. 2: Start of the DESY synchrotron in 1964, left: R. Wideröe
(Photograph: DESY).

We were looking for a good high frequency specialist for the PS project. I knew Dr. Christoph Schmelzer and persuaded him to join CERN. My memory of how this came to be is quite clear. We had agreed to meet near Waldshut in Germany. I came across from Baden and together we drove on to Höchenschwand in the Black Forest. We sat on a nice meadowy slope and I explained the basic principles of the synchrotron to him. He thought that building such a machine could be an interesting thing to do and so he became a member of the PS group.

My official capacity as consultant for CERN came to an end in 1956. After that I occasionally lent a helping hand and later ceased to have any direct contact with CERN. However, I was invited to the congresses (1956 and 1959) on high energy accelerators.

I met Gerry O'Neill [Wi56] during the 1956 congress. He was working on the small storage ring system with colliding beams which I mentioned in chapter 8. He had apparently not heard of my 1943 patent and had developed the principle from scratch. A year later I visited O'Neill in Stanford and explained my War time patent to him. He was quite astonished.

During the period from 1952 to the end of 1959 I attended a total of 19 meetings and congresses about (or at) CERN. I went to most of the congresses of that era.

This is also when I met Ernest O. Lawrence, the inventor of the cyclotron. I think it may have been at the big congress 'Atoms for Peace' at CERN which took place in August 1955. This popular congress would certainly have been a most suitable occasion for a friendly embrace. But it may be that this meeting did not take place until the congress of 1956. Lawrence died of cancer in 1958. I never got to see him in America.

Jan Vaagen told me that there is a picturesque description of the 'Atoms for Peace' (1955) conference in a book by Nuel Phair Davis. It concerns Lawrence who, standing on the podium, with his characteristic sense of drama and pathos, celebrated Professor Wideröe seated in the audience as the author of the basic idea for his cyclotron (in the text it apparently says synchrotron, which is wrong). Lawrence may have had good reasons for doing this, but I can't remember his lecture so well.

My next stint as a consultant was for the research centre DESY in Hamburg between 1959 and 1963. I went there several times and would stay for a few days at a time. Mostly I worked with Dr. Werner Hardt on technical problems related to the construction of a 6.4 GeV electron synchrotron which was planned to have a circumference of about 300 metres. Stan Livingston also spent some time at DESY during that period, but I never saw him. I did meet Gustav-Adolf Voss quite often when the synchrotron was being commissioned. And of course I had many conversations with the founder and director of DESY, Professor Willibald Jentschke. We talked a lot about 'nuclear mills', that is, storage rings with colliding beams, but Jentschke had not been authorized to build such a machine yet.

It wasn't until 1967-68 that collision-machines for electrons and positrons were proposed at DESY and then built and operated – very successfully, if I may add. The first one, called DORIS, was completed in 1974, and the second one, PETRA, went into operation in 1978. It has a circumference of 2.4 km. DORIS is still being used now (1994), albeit for quite a different purpose. With only one beam, it is a dedicated source of synchrotron radiation. Many interesting research and development projects are being performed there.

And then, between 1984 and 1991, HERA was built at DESY. This is a very special machine. The name stands for 'Hadron-Electron-Ring-Anlage' (Hadron Electron Ring Installation) [Wa91]. Electrons (or positrons) of up to 30 GeV are stored in one ring, and protons of up to 820 GeV in another. Both rings were installed in a 6.4 km long underground tunnel. The particles are shot frontally against each other at two points situated within large halls. During one of my Hamburg visits in 1992 Professor Gustav-Adolf Voss, head of the Accelerator Division at DESY, showed me around their impressive installation (see Fig. 12.3). The protons in HERA have to be kept on their course by superconduct-

127

ing magnets. These magnets produce fields which are approximately three times as strong as those of conventional iron magnets fitted with copper coils. A similar type of magnet was also used for a proton-antiproton storage ring called Tevatron, built at Fermilab near Chicago. The Tevatron is about the same size as HERA (6.3 km length) and the particles can be stored at an energy of 900 GeV.

CERN made relatively early use of colliders with their 'Intersecting storage rings' (ISR), a machine in which 30 GeV protons were fired against each other. It went into operation in 1971. This pioneering machine and its use for physics were described by Kjell Johnsen in a CERN-Report with Maurice Jacob [Ja84].

CERN is also where the currently largest storage ring in the world has been built. It is an electron-positron storage ring, which complies exactly with the principle of that patent of mine which was first realised by Bruno Touschek. The ring is called 'Large Electron Positron storage ring' or 'LEP' for short. It has a circumference of 27 km and the particles inside it will achieve up to 100 GeV. In its first phase, when LEP could 'only' manage to bring 50 GeV particles to collide, important work was performed on the Z^0, the neutral exchange particle of the weak force.

I think that LEP is today regarded as the last stage in the development of this type of storage ring. This is because, when storing electrons or positrons in rings, the achievable energy is limited by the synchrotron radiation which is emitted. This is electromagnetic radiation which spans from infrared and visible light up to extremely hard X-rays. The energy lost this way increases drastically with the particle's energy and at a certain point, it is no longer possible (or just too expensive) to replace it even with the most sophisticated means. Only increasing the radius of the machine can help, which again is limited by the costs. This is why it is most improbable that an electron or positron ring with higher energy (or bigger) than LEP will ever be built.

This problem doesn't occur with protons, antiprotons or even heavier particles (at the energies which are available today). These

128

types of particles can be stored in rings with much higher energies, just as long as it is possible to build magnets which are strong enough to keep the particles on their orbits around the ring (as are those used for the Tevatron and for HERA). Current plans for an accelerator in the LEP tunnel at CERN (LHC) appear to indicate the problems which may arise and the limits of the possible. An even bigger project of the same type in the USA has been cancelled for cost reasons.

Thus I have been able to see the storage rings with colliding beams make their triumphant progress through the field of high energy physics. On a personal level, though, I was concerned with quite different issues during that period, stimulated by my acquaintance with medical people for whose work most of the betatrons built at BBC were, after all, intended.

Fig. 12. 3: Rolf Wideröe and Gustav-Adolf Voss in the HERA tunnel in 1992 (Photograph: Pedro Waloschek).

13 How Radiation Kills Cells – the Two-Component Theory

As I was building betatrons it was only natural that I should become more and more interested in their most important application, radiation therapy. By the 60s I was therefore concentrating almost exclusively on the biological effects of radiation, especially in cancer therapy. Until then I had been concerned only with the technology of betatrons. It was a kind of metamorphosis which seemed to me quite logical and, moreover, necessary.

In 1946, when we were designing the first betatron for BBC (the one which was later supplied to the Kantonsspital in Zurich), we already devoted some time to understand better the well known effects of radiation on air and water, especially with regard to the use of electron beams, which we wanted our machines to produce as well. We considered water as a substitute for ordinary cell tissue. This is how we came to select 31 MeV as the most favourable electron energy.

It didn't take us long to discover that both the measuring methods and the units of measurement used were not adequate for beam energies of several MeV (what is now called the 'megavolt region') and that they would have to be updated. These problems became particularly acute later on, towards the end of the 50s, when we extracted high energy electrons from our machines.

Professor Hans Rudolf Schinz and I wrote several papers on this subject. He was in charge of radiation therapy at the Kantonsspital in Zurich and taught at Zurich University. It was not an easy task and sometimes we would have to dig deep into physics in order to get a clear picture of what was going on. It was also difficult to determine the correct radiation doses and we even ended up proposing new units of measurement for them. Professor Schinz performed pioneering work in this field. He made sure that several betatrons were bought in Switzerland and that a great deal of research was conducted in the field of radiation therapy. As a result of his lectures and research work, other countries also installed betatrons for their radiation therapy [Wi59].

The betatron we delivered to Professor Schinz at the Kantonsspital in Zurich was eventually replaced by a new model with higher energy. The older betatron was handed over to the Biological Institute of the University of Zurich which was then directed by Professor Hedi Fritz-Niggli. We made the necessary modifications so that the electrons could be extracted. This betatron is still in use, although Professor Fritz-Niggli retired a while ago. I went to her retirement party which was very nice and she gave a wonderful speech. She also came to my 90th birthday party at the ETH. She and I often discussed the problems of radiation-biology.

The results which were obtained after many years of work have clearly demonstrated that betatrons brought radiation therapy a substantial step forward. I would say that my words at the 1959 International Radiology Congress in Munich were appropriate for their time: "The use of anything other than betatrons for the treatment of deeply situated cancerous tumours should be forbidden by law". Of course I was speaking of X-rays and electrons of up to about 30 MeV energy. However, it took many years for these ideas to spread. Doctors are very conservative people and it is not easy to steer them away from their tried and tested methods. Naturally, there comes a time when they have to accept new findings, but it does cause certain problems for medical research. For example, when we started discussing the new methods of therapy at the Radiumspital in Oslo we were initially regarded almost as charlatans. A lot has changed since then and I would say, albeit with hindsight, that many of the methods which had been used previously caused more harm than good.

Whilst on that same 8th International Congress of Radiologists in 1959 in Munich, I described the therapy of tumours with 31 MeV electrons for the first time and showed that this resulted in a better distribution of radiation dosages than was possible with

X-rays. Irradiation of the affected tissue is improved whilst the rest of the body is subjected to less radiation. A few years later, at the Montreux symposium of 1964, there was extensive discussion of electron therapy and its clinical results. The data reviewed at this symposium was decisive for further development of the irradiation programmes, and it was at this congress that the way forward for high voltage therapy was clarified.

During my years at BBC I had the opportunity to reflect on the correlation of different effects in the irradiation processes. I also had to travel a lot, mainly to give lectures, and in so doing I met many interesting people who were specialised in this field. I stayed in contact with some of them for many years, and my interest grew. This is why I would now like to say something about the physical phenomena which has to be considered in this context.

When fast electrically charged particles (like electrons) penetrate water, tissue or other materials, they generally collide with electrons belonging to the 'electron cloud' of molecules. Thus some of the molecules may end up with one or more electrons missing, i.e. they will have been 'ionised'. This process is therefore called 'ionisation' and it depends on the speed of the particles flying past. Ionisation is higher at lower speeds, which is quite comprehensible since the electrical forces of slower particles have more time to act on the molecules (and their electrons) than do faster ones.

Ionisation processes literally 'put the brakes on' and eventually stop electrically charged particles. Towards the end of their path, the number of remaining ionised molecules increases sharply because the particles travel at slower speed by then. The result is therefore an increasingly dense 'track' of ionised molecules which is left behind by each charged particle at the end of its journey.

However, at the higher energies which we are considering here, an electron (of a molecule) may also receive quite a lot of energy when it is hit, which would cause it to travel a certain distance itself, triggering further ionisation processes. These electrons are called 'delta electrons'. As ionisation greatly increases at the end

132

of the tracks, delta electrons contribute a great deal to the total ionisation effect. And ionisation is the most important factor involved in killing cells. I shall have more to say on this later, especially on the theories developed by myself and others.

First of all though, I must explain a few things about the physical processes which occur when X-rays penetrate matter. X-rays consist of nothing more than high energy light-particles or 'photons'. These can ionise molecules too, by hitting one of their electrons and thus throwing it out of its orbit. At higher energies this is a relatively rare process during which X-ray photons lose a lot of energy and are strongly deviated or even absorbed. Most high energy X-ray photons penetrate through the irradiated body without any interaction. X-ray images are produced by the different rate at which collision processes occur in various substances, which corresponds to different absorptions. Single X-ray photons therefore do not leave a 'track', as would electrically charged particles like electrons.

We end up with rather a complicated picture when we look at the effects of various types of radiation. I have illustrated the most important fact in Fig 13.1 – it is taken from one of my publications on this subject [Wi62].

The top part (a) shows the effect of X-rays produced by a (low energy) 100,000 volts machine on air or water. Ionisation is strongest on the surface and decreases as the X-rays penetrate deeper – the photons are gradually 'absorbed'. A tumour located deep inside the tissue could barely be reached. The surface is subjected to a great deal of radiation and may even suffer burning.

In the centre (b), I show what happens when X-rays of a 30 MeV betatron are used. The radiation is very 'hard', that is, it can penetrate thick layers of matter. Such X-ray photons eventually hit electrons, which can receive a high amount of energy and therefore behave like delta electrons: At the end of their path they cause a lot of ionisation. The radiation effect on the surface is not very strong, which is important, for example, in order to avoid skin damage to the patient.

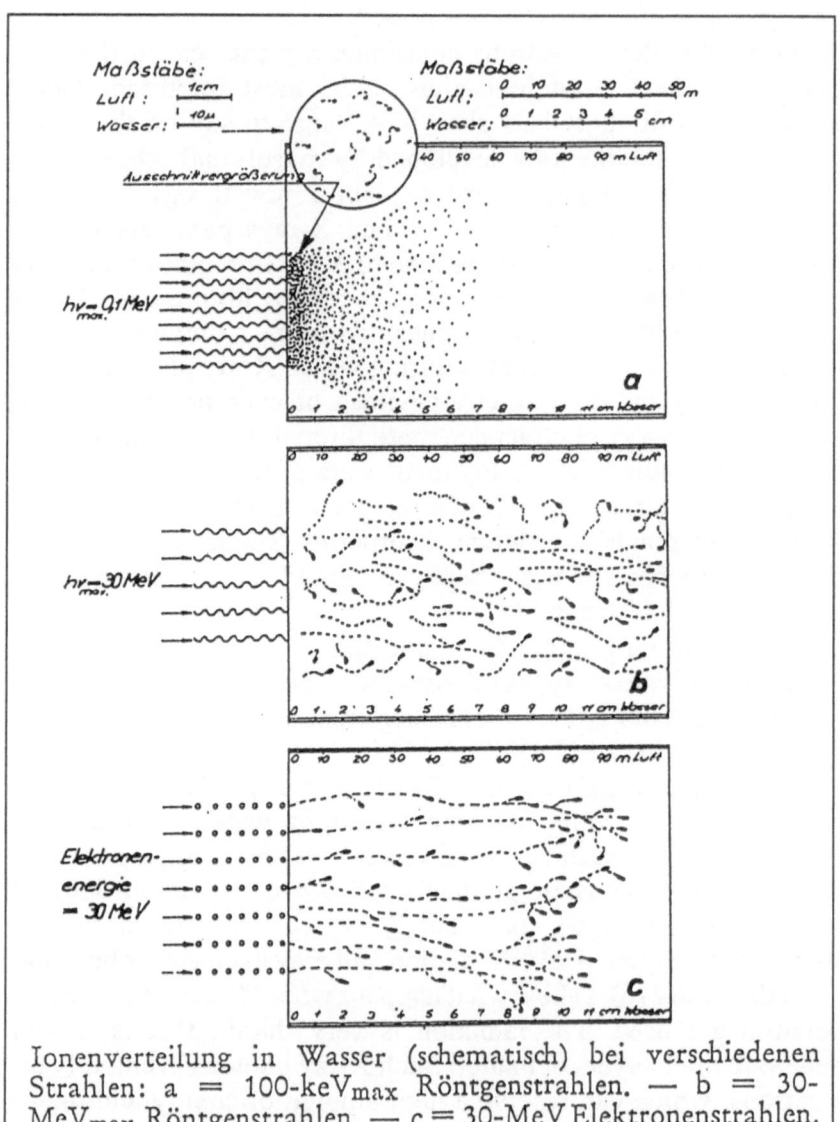

Ionenverteilung in Wasser (schematisch) bei verschiedenen Strahlen: a = 100-keV$_{max}$ Röntgenstrahlen. — b = 30-MeV$_{max}$ Röntgenstrahlen. — c = 30-MeV Elektronenstrahlen.

Fig. 13.1: The effects of X-rays and electrons on matter, as described in the text [Wi62].

And finally, at the bottom of the picture (c), I show how 30 MeV electrons penetrate matter. They ionise on their way (because they are electrically charged), and in doing so lose energy in small stages. In some stronger collisions they also produce delta rays which cause additional ionisation effects. However, the important fact is that the electrons have a limited and defined average range (ionisation and the corresponding energy loss is a statistical process and therefore the 'range' is subject to fluctuations). The region in which most of the electrons 'stop' (where the ionisation effects are strongest) can be determined quite accurately from the energy of the penetrating electrons. The effect on the surface is moderate.

Ionisation of molecules in living cells can have grave, even irreparable consequences. In this context, cancer cells are much more sensitive than healthy cells. Also, healthy cells are better equipped to repair themselves than are cancer cells. This fact is fundamental to the whole of radiation therapy. For example, when a DNA molecule is broken in two places, the almost inevitable result is the death of the cell. During irradiation with alpha rays (for example from radium or other naturally radioactive substances) which have an extremely strong ionising effect, this tends to be the case. Alpha rays are helium nuclei with an electric charge of 2, which move at relatively low velocity and therefore have a correspondingly strong ionising effect on molecules. This is known as the 'alpha effect' – even when it is caused by other types of radiation.

When electrons are used for irradiation, in general just minor, more or less reparable damage occurs to the cells. Only in the worst cases does it lead to the death of a cell. This is called the 'beta effect', named after the 'beta-rays' of radioactive substances which consist of fast electrons.

With regard to the cells which survive following irradiation it is possible to state a formula which I proposed in September 1965 in Rome, unaware of the fact that the same had already been published by M. A. Bender and P. C. Gooch in 1962 [Be62]. I didn't find out about this until 1968. I explained further details and gave references in an article for the periodical 'Strahlentherapie und Onkologie' [Wi90]. The formula is now known as the Bender-Gooch-Wideröe or B.G.W. formula. It provides the probability of survival S of cells following irradiation with a dose D and is made up of two factors, one for the alpha effect and the other for the beta effect of radiation:

$$S = S_\alpha \cdot S_\beta,$$

whereby S_α and S_β are precisely defined functions of the dose D. This might also take into consideration the repopulation effects and properties of various cell types as indicated in Fig. 13.2. This description of the alpha and beta effects of radiation is called the 'two component theory'. It was first formulated by P. Howard-Flanders in 1958 (although without the B.G.W. formula), but received little attention at the time.

In 1960, experiments were already being conducted using various types of radiation on human kidney cells. These experiments proved that alpha and beta effects were independent of each other (G. W. Barendsen [Ba60]). Later on I pointed out that delta electrons (and even further generations of electrons) hitting the cells have to be taken into account additionally.

That is how we finally arrived at a pretty useful picture of the various effects which have to be considered when calculating irradiations. Clinical investigations had also shown that tumour cells react far more sensitively to beta radiation than do normal cells, and this is the main reason for electron therapy providing better results than therapy with radiation containing higher alpha components.

At a meeting of the German Radiology Association – it may have been 1951 in Baden-Baden – I was introduced to Professor Werner Schumacher. After 1960 we met more frequently in Berlin. He was the senior physician in charge of radiation therapy at the Rudolf-Virchow hospital in West Berlin. We had supplied them with a BBC betatron which was inaugurated at the end of

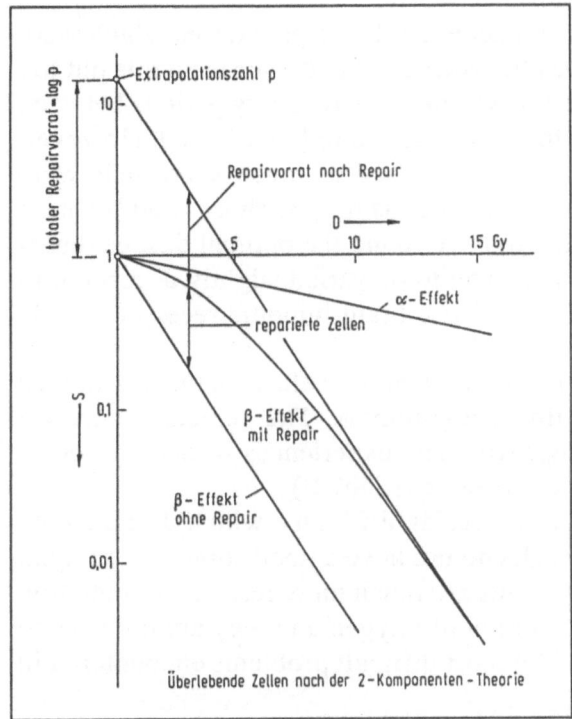

Fig. 13.2: Surviving cells as a function of radiation dosage. The two factors of the two component theory are illustrated [Wi90].

November 1961. It was the first betatron with a 'magnetic lens' and was replaced in 1972 by a 45 MeV Asclepitron. When Schumacher retired in April 1986 I went along to his leaving party and stayed in his house in Berlin. He is at present recovering from a serious traffic accident in September 1993.

Professor Schumacher searched for and tested new and better patient irradiation programmes which were specially adapted for electron therapy of deep-lying lung tumours – his speciality. He dared do many things which other doctors were much less willing to attempt. I worked closely with Schumacher and tried to calculate and explain his results with the help of the two component theory. Our aim was to optimise the electron programmes and to propose a suitable theory. Schumacher irradiated many thousands of patients and gained a great experience in doing so.

In the beginning Schumacher applied single doses which were a little too high (this was between 1962 and 1966) and this put too much stress on the arterial systems. The recovery periods between radiation sessions had to be correspondingly increased. However, when he started to use single doses of electrons which were approximately twice or three times as high as those used by other radiologists, he appeared to have found the optimal dosage distribution. Of course, the dose had to be varied slightly according to the size and type of the tumour – brain tumours received a little more, others perhaps a little less.

In the end Schumacher got far better results than those achieved with traditional radiation programmes. The patients' survival chances were much improved. His experiences went on to be of great use to other doctors (see i.e. [Sch72]).

However, there is a particular difficulty which I shall now recount. Some tumour cells do not have a good supply of oxygen. These so called 'anoxic' cells are much more resistant to radiation than those with a good supply of oxygen and they are not easy to kill. This causes one of the most difficult problems encountered in tumour therapy.

The situation is not entirely without hope however, since radiation changes the supply of oxygen to the tissue. Cells which previously had too little oxygen, start to take in more and can thus be killed during the subsequent radiation session. However, this causes a considerable uncertainty factor which affects both calculation and therapy.

It was a great step forward then, when, in the period between 1973 and 1986, Professor Wolfgang Pohlit discovered a new way to improve the killing of tumour cells and in particular those with an inadequate supply of oxygen [Pu82]. Pohlit's weapon was to treat the patient with 2-deoxy-D-glucose (2-DG). This substance is so similar to ordinary glucose that the tumour cells (especially those which lack oxygen) absorb it. However, the 2-DG blocks the glucose and therefore undermines the energy sources of the oxygen deficient cells and so they die off quite quickly. 2-DG does

138

not appear to have any harmful effects, and the first clinical tests proved positive. I believe that this removed a major uncertainty in radiation therapy and I would count it amongst the greatest advances of recent times.

But let's get back to Schumacher's radiation therapy in Berlin. The fact that he was able to use higher single doses with electrons is easily explained. Electrons have the lowest alpha effect of all types of radiation. Consequently the total radiation effect at the usual dose values is correspondingly low. It is therefore necessary to apply higher single doses in order to achieve the same radiation effect. At the same time this avoids killing normal cells with alpha effects. The optimal single dosages and radiation programmes which Schumacher arrived at could probably be improved upon by means of Pohlit's 2-DG therapy.

It was not at all easy to overcome the orthodoxy of some of the surgeons in this field. I can clearly remember what took place at the Radiumspital in Oslo. I had recommended that Dr. Rennäs, who worked there, should visit Schumacher and had arranged an appointment. When the time came, Dr. Rennäs wrote to me (he has since passed away) that his director had strictly forbidden him to go to Berlin and visit Schumacher. The traditionally orientated senior surgeon was obviously somewhat fearful of the newer methods.

Metastasis is still a major problem for radiation therapy. Many experiments have been conducted using poisonous substances to kill cells, but the results have been more than merely doubtful. New ways are now being tried, such as utilising the immune system to dissolve tumours.

I met a very interesting man at the Radiation Research Congress in Evian, I think it was in 1970. This was Dr. Lionel Cohen. I had the opportunity to have some longer talks with him during two subsequent visits to Johannesburg in South Africa where he was leading radiation therapy in a big hospital. We stayed in contact for many years. Cohen is an excellent radiologist and has had many good ideas. He moved on to Chicago (USA) and has since retired.

Box 14

The Success of the Megavolt Therapy

Wideröe has clearly shown the advantages of using photons and electrons of higher energy for radiation-treatments. Betatrons were first used for this purpose after the Second World War. Due to their compactness, they could be mounted in a suspended position or even be mobile. They were built for energies of up to 45 MeV. Cobalt 60 bombs were also used for radiation therapy during that period. After 1970, compact and relatively economical linacs increasingly came into use. Their technology was very reliable and had its origins in particle physics laboratories. The most important developments were achieved in the Stanford Linear Accelerator Centre (SLAC) in the USA.

The physicist John Ford, an expert in this field and Vice President of 'Varian Health Care Systems', reported in 1993 [Fo93] that approximately 3,500 linacs were being used for radiation therapy throughout the World, half of them in the USA. Usually these linacs reach an electron energy of about 20 MeV. Higher electron energies (30 to 45 MeV, as it was the case with betatrons) are rarely used today.

According to John Ford, more than half of all cancer patients (in the USA and Western Europe) are today treated with radiation therapy, used either as the main form of therapy or in conjunction with surgery and chemotherapy. The most important fact about this is that in about 50% of the cases which are pronounced healed, the cure can be entirely or partly attributed to radiation therapy.

Today, electron irradiation is used in 10 to 15% of all therapy cases, the remainder use X-rays, whereby technical progress in the equipment has improved the irradiation quality. The linacs used for this are relatively small and reach up to 20 MeV within 60 cm of sophisticated iris-loaded wave-guides in which an electromagnetic travelling wave accelerates the electrons.

Cohen had confirmed the correctness of Schumacher's programmes for electron therapy with higher single doses and increased recovery periods. He seemed to place particular importance on the fact that tumour cells mend at a much slower rate and much less successfully following beta damage than do normal, damaged cells. The same applies to the repopulation of dead tissue. However, this subject has not been deeply investigated so far. Cohen recognized the decisive importance of the parameters in the B.G.W. formula and set everything on deriving these from the practice of radiation therapy. He soon extended the two component theory by a third component; a very interesting development. He took into consideration the destroyed cell tissue and was thereby able to come up with even better programmes for irradiation. This difficult task could only be possible with the help of computers and he developed the required software which he described, together with his methods, in a book published in 1983 [Co83].

I also had a very good relationship with the chief surgeon of a hospital in Beijing where we had installed a BBC betatron. I had to go to China two times to give lectures. Of course, during these lectures I explained precisely how a betatron is made up and how it works. On a subsequent visit I discovered that the Chinese had built their own betatron in the meantime, which complied exactly with the contents of my lectures. It worked rather well too, except that they were not able to extract the electrons, something which was possible with our betatron.

In many cases, electron therapy has proven to be an improvement on X-ray therapy. The 'magnetic lens' for electrons mentioned earlier, which we developed at BBC, also came to be applied. It was made up of rotating permanent magnets which bent the electron beam and thus steered it towards the object to be irradiated from continually changing directions. This distributes the stress on the tissue layers above even more advantageously. Use of the 'lens' was essential to gain full profit from extracting the electrons from the betatrons.

When I first started working in this field nobody really knew anything precise about the primary physical effects of radiation. Nowadays we know, for instance, that the secondary electrons, the delta electrons, have a major role to play. The next effect which should be investigated is purely biological: the effects of delta electrons on enzymes, and in particular, on the DNA molecule.

And with this we have come directly to the big question: How are cancer cells created? We believe today that we know something about this. They come about by means of certain enzyme mutations. But the problems and the possibilities are various and the scientists are still a long way from concluding their research.

Naturally there are many more details on which I would like to expand, but I think it may be better to stop here. For further questions on cancer therapy and the application of betatrons I would like to refer to the many articles I have written on the subject (see i.e. [Wi90]).

My occupation with the uses of the betatron, especially in the field of medicine, required me to pay many visits to the institutes and hospitals to which we had supplied our machines. Of course, I also attended at as many conferences and congresses on the subject as I could to keep myself well informed of the latest developments. The list of my trips after 1947 is very long and quite interesting. I always wrote down the purpose for which I was making a trip (a conference or perhaps a lecture – sometimes it was several lectures), and the names of the most important people I met there. This helps to refresh many memories, for instance the two beautiful dresses I brought my wife from Beijing and my visits to the Krüger animal park in South Africa.

And last but not least, I became the recipient of many honours as a result of my work in the field of radiation biology. In fact, they were more numerous than for my developments and ideas on particle accelerators. This may have a lot to do with the lectures I gave all around and with many articles which I published on radiation therapy.

14 Some Retrospectives and Dreams

As I speak about my life, I find that I return frequently to a few very special events which I now consider to be the most important stages of my work. While I was actually involved with these things I wasn't really aware of their relevance or future importance, since everything I did generally gave me much pleasure and I always concentrated completely on whatever I was doing. Thus I built relays with just as much enthusiasm as I later constructed betatrons. And I was always particularly interested and motivated if there were new ideas involved.

However, what always comes to my mind first is the Aachen drift-tube. Proving that it was possible to accelerate electrically charged particles with alternating potentials and without having to use the restricted possibilities of the (at that time, usual) d.c. voltage, appears to me as my most fundamental piece of work. This was the major result which I presented in my dissertation in 1927 and it does appear to have had the most far-reaching consequences. Added to this was the happy circumstance that this work was widely disseminated and well known everywhere. It is definitely one of the most quoted publications on particle accelerators.

The 'bent drift-tube' appeared first in Lawrence's cyclotron and later in the accelerating cavities of the synchrotron. The latter now seems much more important to me because the synchrotron formed the basis of storage rings. My discovery of the stabilized particle orbits in synchrotrons might also have been quite important. However, the further development of the drift-tube which took place at almost the same time as the cyclotron, starting with Alvarez' resonators, via the cavities with standing waves and finally resulting in the iris-loaded wave guides with travelling waves of modern linear accelerators, is also very interesting. All this began in 1927 with the first drift-tube in Aachen.

The 1943 patent containing my invention of storage rings [Wi43a] was probably very important but it was kept secret for ten years. As I could not see any practical use for it myself (there were still too many technical problems which needed solving), I did not speak much about it. I was not to explain my proposals again until the 1956 accelerator conference in Geneva [Wi56], after Kerst and O'Neill had rediscovered the principle. However, others took the development further while I was fully occupied building beta-trons. I am therefore very pleased that I had had the right idea thirteen years before my colleagues, but I can't blame them if they've sometimes forgotten me, since they would have often spent years working on these projects. Many very beautiful storage rings were built while I was busy with other problems.

I think it is pretty clear from my story that I was deeply committed to my work with relays. I guess my contributions to this field were quite good and I believe that my relays were also very useful. Although it might not be of great interest to particle physicists and doctors, this work was creative and I am rather proud of it.

I endow my work in the field of radiation therapy with a certain amount of status. It is on this subject that I had the opportunity, for the first time in my life, to be active as a scientist at a highly regarded institution, the ETH. This was a completely new experience for me and I was able to let my imagination run free without having to take into consideration the interests of an industrial company. It must be said however that BBC, who were still employing me, were not at all opposed to my lecturing, because I was in some way contributing to the sale of betatrons. I hardly published any technical articles or applied for patents during this period, but concentrated instead on writing for scientific periodicals and giving lectures.

My increasing interest in radiation therapy was a logical continuation of the war against the tumour cells with our new weapon: the megavolt beam. After all, the patients needed urgent help and I took part in this with a great deal of enthusiasm.

However, while I was busy with all these other things I never lost sight of particle accelerators. I kept up to date by reading periodicals and speaking with my many friends.

That is how I came to follow the exciting development of cyclotrons while still in Berlin, through the news which Ernst Sommerfeld used to bring to me from his father. Of course the situation was rather more difficult during the War, but from about the end of the 1940s onwards a completely new scientific spirit took over. Communication between scientists became desirable. Unrestricted travel, mutual visits and international conferences meant that people knew just about everything that was happening in their field. One even knew most of the participants on a personal level, which was essential for the impressive progress in the field of particle physics and the structure of the smallest particles of matter.

Nowadays it is easy to keep quite well informed on many areas of research, as long as one has enough spare time for reading – and a few good friends. So, even after my retirement, I could not refrain from studying the basic problems of particle acceleration. Only through experiments at even higher energies will we be able to obtain new knowledge which should finally lead us to a comprehensive theory of the structure of all kinds of matter.

Well, after the successful eras of cyclotrons, synchrotrons and now storage rings, we have gone back to basics: Experts agree that probably there will be no bigger rings in future and that linear accelerators will be built instead. I have already mentioned the reasons for this: Electron and positron rings are limited by their synchrotron radiation, and proton rings are disadvantaged by their need for stronger magnets and by the cost of gigantic rings. After all, plans can only be made for those accelerators which can realistically be built with the means available, and obviously, these means are limited.

Ideas are not subject to any such considerations. The limitations are set only by the intellect of human beings themselves. The theoretical possibilities with regard to accelerating particles by

145

Box 15

Wideröe's Life at a Glance

1902	**Born in Oslo**
1922	Karlsruhe: **Betatron idea**
1927	Aachen: **First linac works**
1929	Lawrence: First 80 keV cyclotron in operation
1929...	Berlin: **Construction of distance relays**
1933...	Oslo: **Construction of distance relays**
1941	Kerst: First betatron (2.3 MeV) in operation
1943	Oslo: **Storage ring idea, patent**
1944	Hamburg: **15 MeV betatron works**
1945	McMillan, Veksler: Synchrotron
1945	Oslo: **Synchrotron theory, patent**
1946...	Baden: **Construction of betatrons at BBC**
1952...	Synchrotrons: Cosmotron, Bevatron, PS...
1952...	Geneva: **Consultant at CERN (PS project)**
1953...	Zurich: **Lecturer at ETH Zurich**
1956	Kerst and O'Neill: Re-invention of storage rings
1956...	Baden: **Construction of the Turin synchrotron**
1959...	Hamburg: **Consultant at DESY (synchrotron)**
1959...	Baden: **Megavolt radiation therapy**
1960	Frascati: Touschek, AdA, first storage ring
1962	Aachen: **Dr. honoris causa at RWTH Aachen**
1963...	Triumphant progress of storage rings
1964	Zurich: **Dr. med. h. c. at Zurich University**
1965...	Baden: **Two component theory**
1969	Remscheid: **Röntgen Medal**
1971	Würzburg: **Röntgen prize**
1973	Oslo: **Member of the Norw. Acad. of Science**
1973	Madrid: **JRC gold medal**
1992	Washington: **Robert R. Wilson Prize of APS**

electromagnetic means (i.e. within the scope of the Maxwell equations which have been known since the 19th century), are nowhere near being exhausted, and technology surprises us almost daily with innovations which in turn allow us to broach new trains of thought. Although many of the ideas in this field which appeared over the last decades were not successful, it is possible, in principle, that there are yet more fundamental breakthroughs to be made. They could allow us to advance to energies unimaginable today. We have to remember that the things we build today appeared utterly utopian 50 years ago.

I would like to mention such an alternative as a vision of the future, not because I am fully convinced that it is good or correct, but because I consider it important that we maintain our confidence in further developments, however adventurous they may appear.

This story begins in 1956 at the International Conference on Accelerators in Geneva where Veksler presented a report on a very peculiar idea which rather impressed me. A fast bunch of particles was to be made to 'meet' or 'overtake' a slower bunch of other particles and thus 'sweep it along' in its path. He indicated a number of possibilities. As some of Veksler's statements did not seem quite right to me, I thought the matter over and wrote down my results in April 1986. Veksler had christened his methods 'coherent acceleration'. This name is apt since the particle bunches have to act on each other as entireties, i.e. they must be 'coherent', and the individual particles must not interact. During ordinary acceleration we look at individual particles. We do not consider effects which affect the entire bunch until later, when the orbits are being corrected – not during the acceleration process itself.

I had come to think that it would be best if a bunch of protons could be 'hit' from behind by a bunch of positrons (10^4 positrons for each proton), and in my considerations I just took as an example the data of the particle bunches which could be available at the HERA rings in Hamburg, i.e. 800 GeV protons and 30 GeV positrons. Measured in the rest frame, the positrons will have an energy of 17.1 MeV. The results are rather startling. It becomes possible to accelerate the proton bunches so that each proton has an energy of several hundred TeV, and under the best conditions even over one thousand TeV. It is therefore possible to achieve extremely high energies by coherent scattering of particle bunches. In comparison, the protons in the LHC storage ring proposed at CERN would reach no more than 8 TeV.

For my 1986 calculations (and an addendum I made with J. F. Crawford) I had to assume certain bunch sizes and I also mentioned many of the difficulties that may be expected (We never published these considerations – I just sent a few copies to my friends). The major factor for the realisation of this method is the size of the particle bunches. I used the actual dimensions of the HERA bunches, which are several centimetres long, a few millimetres wide and only a couple of tenth of a millimetres high. However, the coherent scattering principle would work much better if it were possible to make the bunches much smaller. At the time I made my calculations, this was still thought to be unrealistic.

And this is really where the point of my story lies. In 1992 I was informed of the new plans for linear accelerators to be built in the future, since storage rings have now reached their limits. Much higher collision energies should be achieved in future, but as a first goal, electrons and positrons would be shot against each other using two linacs, each one providing particles with an energy of just a few hundred GeV, a value not at all accessible to storage rings for this type of particles.

The things that interested me most however, were the dimensions of the particle bunches in these linacs. I was very surprised when I heard that the aim was to have bunches which were about a factor one thousand smaller than those available with today's machines. This is the only way to achieve reasonable collision rates – a conclusion I had already reached in 1943, when I invented the principle of storage rings with colliding beams – just to overcome this difficulty. If in the past we have considered a few tenths of a millimetre as possible transverse beam dimensions, we are now talking about tenths of micrometres. For some projects, people are even speaking of hundredths of micrometres, which is the same as ten nanometres. The particle bunches which are going to interact coherently will have to be localised in space and steered with even greater precision. When this precision is achieved, it will perhaps be possible to think of other mechanisms, apart from 'coherent bunch collisions', with which to accelerate particles to extremely high energies.

However complicated and utopian all this may seem to us now, it would undoubtedly be of great interest for physics research, if protons with 1,000 TeV were available. Today, this kind of energy can only be found in cosmic radiation, that is, in particles arriving from intergalactic space – and then, only very rarely.

It would be easy to come to the conclusion that the builders of accelerators who follow such fantastic ideas were completely mad, if we had not all been party to the developments of the last

---

Box 16

Wideröe's Memberships:

American Physical Society
American Radium Society
British Institute of Radiology
Deutsche Röntgengesellschaft (honourary)
European Society for Radiation Therapy ESTRO (honourary)
European Society of Physics
Naturforschende Gesellschaft (Zurich)
Norwegian Society of Radiology
Norwegian Society of Physics
Schweizerische Physikalische Gesellschaft (honourary)
Schweizerische Gesellschaft für Radiobiologie (honourary)
Scandinavian Society for Medical Physics (honourary)
Society of Nuclear Medicine

decades: A few years ago no technically versed person would have believed that the precision which is now used in the production of millions of CD-disks' would ever be possible. This example shows that we should never lose courage and that we must continue to aim for goals which lie far beyond us, even if they are still absolutely held to be at times unattainable.

With this I shall end this story of my life, but not before I have thanked the readers for having made it this far and for their interest and patience.

Chronological Survey

Entries emphasized in *italics* refer directly to the life and work of Rolf Wideröe *(RW)*. I have included events, which could have had relevance for Wideröe's life. They were originally collected as editing aids and I make no claims for completeness.

P.W.

Year-Month-Day	
1902-07-11	*RW born in Oslo.*
1905	Nobel prize to Philipp Lenard.
1906	Nobel prize to Joseph J. Thomson (the electron).
1911	Ernest Rutherford discovers the atomic nucleus.
1918	Nobel prize to Max Planck (quanta).
1918	Rutherford: first disintegration of an atomic nucleus.
1920-summer	*RW's school-laving exams at the Halling School in Oslo.*
1920-autumn	*RW begins studies in electrical engineering at Karlsruhe Technical University.*
1921	Nobel prize to Albert Einstein.
1922	Nobel prize to Niels Bohr.
1922-01	1 $ = 192 German Mark.
1922-04-01	J. Slepian (Westinghouse) applies for a US-Patent 'X-Ray Tube' presenting the first rudimental ideas for a betatron [Sl27]; published on Oct. 11, 1927.
1923-03-15	*RW's first (preserved) notes in a copy-book including a sketch for a betatron [Wi23]. (More drawings and computations in other copy-books.)*
1923	*RW's one-month practical work in a factory for electric motors in Strasbourg.*
1923?	*RW asks an agency in Karlsruhe to submit a patent on the betatron. It was probably never submitted. (The agency's building was completely destroyed during the War.)*
1923-11-15	1 $ = 4,200,000,000,000 German Mark.
1924-03-12	G. Ising: First known proposal for the acceleration of charged particles with electromagnetic 'travelling waves' [Is24].

1924	*RW obtains his diploma in electrical engineering at Karlsruhe Technical University. His thesis was on 'Voltage Distributions in Chains of Isolators'.*
1924	*RW's first publication, on 'Inflation in Germany' [Wi24].*
1925-summer	*RW's 'practical work' in the locomotive factory of the Norwegian State-Railways. He also completes 72 days of military service in Norway.*
1925-autumn	*RW proposes the 'ray-transformer' (betatron) as thesis for a doctor-degree in electrical engineering in Karlsruhe. Prof. Schleiermacher (theory) agrees, Prof. Gaede (physics) refuses. Gaede assumes that the achievable vacuum would not be sufficient (residual gas would absorb the circulating particles).*
1925-End	*RW studies Lenard's publications [Le18] on the absorption of electrons in matter and comes to the conclusion that Gaede's assumptions were wrong.*
1926-05	*RW proposes the construction of the ray-transformer to Prof. Rogowski in Aachen.*
1926-06...	*RW starts working and studying at the Technical University in Aachen (RWTH) under Prof. Rogowski. Tests of the first ray-transformer (betatron) are unsuccessful due to surface charges in the tube and lack of stabilizing forces of the magnetic steering field.*
1927-autumn	*RW changes over to building a small linear accelerator. He succeeds in accelerating ions to 50,000 volts, having only 25,000 volts at his disposal. It is the first drift-tube ever operated, demonstrating the principle of acceleration of charged particles with high frequency alternating voltages.*
1927-autumn	Steenbeck starts working with Rüdenberg at Siemens Halske company in Berlin.
1927-10-11	Slepian's US-Patent (Westinghouse) 'X-Ray Tube' is made public [Sl22].

1927-11-28	*RW finishes all examinations and obtains his 'Dr.-Ing.'-degree in Aachen. The successful linac with one drift-tube is the main subject of his thesis, the ray-transformer (betatron) is explained in Section IV including the '2:1-ratio' between accelerating and steering fields, which is later called the 'Wideröe-relation' for betatrons.*
1927	Breit and Tuve (Carnegie Institution USA) perform interesting tests with a simple betatron [Br27]. Their efforts are unsuccessful, but very promising.
1928	*RW's dissertation is published in the 'Archiv für Elektrotechnik' [Wi28].*
1928-03	*RW moves to Berlin. He obtains a position at AEG's transformer factory (Berlin Oberschöneweide) following a recommendation from Rogowski. RW develops safety-relays for short circuits in power lines. By the end of 1932 he has applied for 42 German patents and 2 US patents, all for AEG.*
1929	Walton reports on tests of a simple betatron and a linac built at Cambridge, following suggestions by Rutherford. None of these devices work. However, Walton includes very important deductions and formulas in his publication, establishing for the first time precise stability conditions for circular orbits in betatrons [Wa29].
1930	Breit, Tuve, Hafstad and Dahl develop several very interesting high voltage generators at the Carnegie Institution in Washington DC.
1930	Lawrence and Edlefsen publish the basic ideas for a 'cyclotron' [La30].
1931-01	Lawrence communicates the successful operation of his first cyclotron (13 cm diameter, 80 keV) to the American Physical Society [La31b].
1931	Lawrence and Sloan construct and operate a linac following Wideröe's ideas. It has 15 drift-tubes and reaches 1,26 MV [La31a]. Other linacs follow.

Year-Month-Day

1931	Van de Graaff communicates to the American Physical Society the successful operation of his first electrostatic generator using a silk-band [Gr31] with which he achieves about 1,5 MV. Several similar installations follow.
1932	Cockroft and Walton [Co32] succeed in obtaining the first nuclear disintegrations using artificially accelerated particles (400 keV cascade generator). Lawrence confirms this results a few months later with a 1.2-MeV-cyclotron.
1932-12	Lawrence successfully operates a 69-cm-cyclotron for 4,8 MeV.
1932-12	*RW moves from Berlin to Oslo, scared of the economic crisis in Germany and of Hitler's rise to power.*
1933-03-01	Rüdenberg and Steenbeck (Siemens-Schuckert-Werke, Berlin) apply for a German patent [Ru33] which includes a rough stability condition for a betatron (published on Febr. 4, 1938). As is usual at that time, no references to previous work are given. When the patent is submitted Rüdenberg has already emigrated to Great Britain to escape anti-semitism.
1933	Ising publishes an article in the Annual-Report of the Swedish Physical Society [Is33] in which Wideröe is wrongly described as 'German' (page 34).
1933-04-01...	*RW builds protective relays for the company N. Jacobsen in Oslo. By 1937 he has applied for ten Norwegian patents on relays.*
1933-autumn	*RW's driving holiday in a Ford-A. From England (with his friend Torwald Torgersen), to France, Spain and Germany. RW also tries to sell his relays. He has no success and experiences severe health problems.*
1934-02	*RW meets Ragnhild Christiansen (born Jan. 3, 1913), in Ms. Fearnley's dance academy in Oslo.*
1934-11-14	*RW and Ragnhild are married.*

1935-03-07	Steenbeck (Siemens) applies for a second patent in Germany (also in Austria) for a betatron [St35]. Besides the rough stability condition, this patent also includes (as claim) the 2:1-relation between steering and accelerating field.
1935-Middle	*Ragnhild Wideröe works for a short time (unofficially) at Jacobsen's and helps RW build and test relays.*
1936	Jassinski: interesting paper on betatrons [Ja36].
1936-03-06	Steenbeck (Siemens) applies for a betatron-patent in the USA [St36] (published on Dec. 28, 1937).
1936-06-25	*RW's daughter Unn is born in Oslo.*
1937	*RW's chance discovery of Slepian's US patent [Sl27].*
1937 ??	*RW's report on relays in Copenhagen (Nordish Engineer's meeting). Ing. Styff (from NEBB) is present.*
1937-04...	*RW starts working for the transformer factory 'National Industri', in Oslo, a subsidiary of Westinghous USA. Very boring activity!*
1938-12-20	*RW's son Arild is born in Oslo.*
1937-12-28	Steenbeck's US-Patent on betatrons is published [St35].
1938-autumn	The 'Physics Association' is founded in Oslo.
1939-summer	First edition of the Norwegian review 'Fra Fysikkens Verden' published by the Physics Association.
1939-09-01	German troops invade Poland. Great Britain and France declare war against Germany.
1939-10	Lawrence operates his 150-cm-cyclotron for 19-MeV deuterons.
1939-11	Nobel prize to Lawrence.
1940-04-09	German troops occupy Norway.
1940-05	Touschek is expelled from Vienna University as 'non Aryan'. Takes several jobs. He helps Arnold Sommerfeld revise Vol. 2 of the famous book 'Atombau und Spektrallinien'.

1940-06...	*RW starts working for 'Norsk elektrisk og Brown Boveri' (NEBB) in Oslo, planning and building power plants.*
1940-10-15	Kerst (Univ. of Illinois) reports on successful tests of a 2 MeV betatron [Ke40a]; Wideröe and Walton are quoted, Steenbeck is not.
1940-11-13	Kerst ('on leave at General Electric') applies for a US-patent for a betatron [Ke40b].
1940-11-22	Kerst ('on leave at General Electric') reports on the successful operation of his 2,3-MeV-betatron [Ke40a].
1940-End	Touschek goes to Hamburg, R&D at the 'Studiengesellschaft für Elektronengeräte' (Philips). He is allowed to hear (illegally) lectures by Professors Lenz und Jensen at the University of Hamburg.
1941?	General Electric asks Siemens for a licence to use Steenbeck's betatron patent [Ka47] [St77].
1941-04-18	Kerst (General Electric) submits his famous paper on the operation of the 2,3-MeV-betatron [Ke41a] to Phys. Rev. He reports on gamma rays equivalent to about 1 gr of radium. Wideröe's, Walton's und Jassinski's papers are referred to (according to RW, on a request by the editor), Steenbeck's patent is not. In a subsequent paper Kerst and Serber describe the corresponding theory [Ke41b]. According to Professor W. Paul, this is the last issue of the Phys. Rev. which arrives legally in Germany; it arrives illegally in occupied Norway (Trondheim), mailed as an ordinary letter.
1941-09-03	*RW's son Rolf is born in Oslo.*
1941-autumn	*RW hears Roald Tangen's seminar at the 'Physics Association' in Oslo, in which he reports on the Kerst-betatron. RW realizes that it is possible to construct a ray-transformer and starts working on the subject again.*
1941-12-06	According to Max Steenbeck [St77], Siemens licensed General Electric to use his patents the day before the Japanese attack on Pearl Harbor.

1941 End	Konrad Gund (X-ray engineer) begins planning a 6-MeV-betatron (550 Hz) for medical purposes at the 'Siemens Reiniger Werke' in Erlangen, prompted by Steenbeck.
1942-02	Steenbeck's publication in 'Electronics', February issue 1942, pp. 22-23.
1942-??	RW's brother Viggo (born 1904, a pioneer of Nowegian air transport) is imprisoned in Germany after trying to help resistance members to escape from Norway to England.
1942-07	According to Kaiser [Ka47], Siemens applies for a betatron patent, 'Akt. 151 465, VIII c/211g'.
1942	Kerst reports on the operation of a 20-MeV-betatron and introduces the name 'betatron' [Ke42].
1942-09-15	*RW submits a paper on betatrons to 'Archiv für Elektrotechnik' [Wi43b], describing his own and Kerst's work, as well as some new ideas on betatrons from 10 to 1000 MeV and a detailed design for a 100 MeV betatron, including cost estimates.*
1942-09-29	The US-patent 'betatron' of Kerst [Ke40] is published.
1942-End	Touschek moves to Berlin, works at 'Opta Radio' on the development of Braun-tubes, the predecessors of the klystron-valves used later on for radar applications. *He also works for Dr. Egerer, the editor of the 'Archiv für Elektrotechnik' and sees Wideröe's proposal for a betatron. He finds a mistake in the relativistic calculations and writes to RW, who asks him to join him (quoted in [Am81] p. 5). (RW can not remember these letters; there seem to be no copies preserved.)*
1942-12-15	Steenbeck, Dr. Kurt Bischoff, Dr. J. Patzeld und (Dr.) Konrad Gund: meeting on the new betatron project, following ideas of Jassinski [Ja36], quoted in [Ka47].
1943-01-31	Capitulation of German troops in Stalingrad.

1943-spring | *Visit of (2 or 3) German Air Force officers to RW at NEBB in Oslo. Two days later RW is taken to Berlin by air. It is implied that they would help get his brother Viggo out of prison. According to RW, this is why he accepts to go to Germany. He is to build first a small betatron for 15 MeV in Hamburg and some larger ones later on. This is to be done as 'compulsory labour' with the agreement of NEBB Company (BBC).*

1943-05-08 | Prof. Jensen discusses with Schmellenmeier plans to build a 1,5-MeV-'Rheotron' (Jensen had previously agreed this with Prof. F. Houtermans).

1943 | Steenbeck reports in 'Naturwissenschaften' on a 1,8 MeV betatron (a secret project at Siemens) which had already been in operation in 1935/36 and explains his early ideas and patents on betatrons [St43].

1943-07-12 | *RW submits a second article on betatrons to 'Archiv für Elektrotechnik', which includes ideas for a 200-MeV-machine. It is not published.*

1943-07-15 | *RW applies for his first patent on betatrons in Germany on 'Injection' (No. 889659), accepted on Jul. 30, 1953, published on Sept. 14, 1953. RW receives legal advice from his friend Dr. Ernst Sommerfeld (Berlin), the son of Arnold Sommerfeld, for all his German patents .*

1943-07-25 | (to 1943-08-04) Operation 'Gomorrha': Five allied bombings on Hamburg cause great destruction. *During these days RW is not in Hamburg.*

1943-08-05 | The 'Reichsforschungsrat' (German Research Council) orders a 'Rheotron' (betatron) from Schmellenmeyer (Berlin) [Sw92].

1943-08... | *RW starts working in Hamburg; rents a room. Back and forth between Hamburg and Oslo. Occasional visits to Berlin. His family remains in Oslo. His salary is paid to his wife in Oslo. RW gets in touch with Hollnack and Richard Seifert (trustees of the German Aviation Ministry) und with physicist Dr. Kollath.*

1943-?   *RW meets Bruno Touschek for the first time at the home of Prof. Lenz. Touschek starts working with RW, makes theoretical calculations for the betatron, i.e. on radiation losses (also for an already envisaged 200-MeV-machine) and orbit studies, using the Hamilton formalism.*

1943-08-End   *RW takes a vacation in Tuddal near Telemarken (Southern Norway) and has the idea for 'storage rings' whilst lying on the lown behind his hotel. These are expected to provide higher energy and improved collision rates for nuclear reactions.*

1943-09-02   *RW applies for a 2nd betatron-patent in Germany on 'electrical lenses', No. 927590, published Dec. 5, 1953.*

1943-09-02   *RW applies for a 3rd betatron-patent in Germany on 'premagnetisation', No. 932194, published Aug. 25, 1953.*

1943-09-04   *RW applies for a 4th betatron-patent in Germany on 'opposite magnetisation', No. 925004, published on March 10, 1955.*

1943-09-08   *RW applies for a German patent on 'storage ring collider'; No. 876279 [Wi43a], published on May 11, 1953.*

1943-09   *RW meets the editor of 'Archiv für Elektrotechnik', Dr. Egerer, as well as Dr. Schiebold (the physicist promoting 'death-X-rays' for shooting down aeroplanes) at Hollnack's.*

1943-10-01   *RW's report (unpublished) on the development of betatrons, including many references [Wi43b].*

1943-10-05   *RW applies for a 5th betatron-patent in Germany, on 'magnetic lenses', No. 932081, published Nov. 10, 1955 (addendum to patent No. 927,590)*

1943-11   *Begin of design and construction of a 15-MeV-betatron at C. H. F. Müller-company (Philips) in Hamburg. Iron plates supplied by Seifert's factory, cathodes by Boersch. There is a detailed report with drawings by 'Dr.Müller' at the ETH-Library [Mu43].*

1943-11-06	*RW proposes (report from Oslo) a 'fast schedule' for building betatrons in Germany. It includes: a) a 15-MeV-betatron in Hamburg, b) a 200-MeV-betatron and c) a future test-laboratory in Groß-Ostheim. He mentions that work on the 15-MeV-machine has already been started at C. H. F. Müller company in Hamburg [Wi43c].*
1944	*RW applies for five further betatron patents in Germany.*
1944-04	Gund's 5-MeV-betatron is successfully operated for the first time at Siemens in Erlangen. Machine parameters are measured and first experiments are performed (in Erlangen) by H. Kopfermann and W. Paul (both from Göttingen). *RW is apparently unaware of these activities.*
1944-04-27	*(to 1944-04-29) Visit to BBC in Weinheim. Minutes by RW dated May 1, 1944 [Wi44]: Meyer-Delius (BBC) reports on Bothe and Gentner constructing a betatron with extracted beam. (Gentner perhaps confused with Dänzer, who planned with Bothe a betatron for 10 MeV, as reported by W. Paul [Pa47], p. 51.)*
1944-04-29	*'Secret' minutes by Meyer-Delius on a BBC-meeting in Heidelberg (present: Seif(f)ert, Wideröe, Meyer-Delius, Kade, Weiss, Kneller) to discuss the construction of a large betatron following the 'megavolt procedure' [Me44]. Seif(f)ert had passed a 'provisional order' from the German Aviation Ministry for BBC to start R&D for such a machine.*
1944-06-13	Start of V1 flying-bomb attacks on London.
1944-summer	*The 15-MeV-betatron is successfully operated for the first time in Hamburg.*
1944-08	According to Kaiser [Ka47], there is a 'contract with BBC Heidelberg to built a 200-MeV-betatron' (?). *(According to RW there was no BBC representation in Heidelberg at that time. One of the directors lived there, and a few meetings were held in Heidelberg; see [Me44]).*
1944...	Touschek writes several reports on the theory of betatrons. Some of them are preserved at the ETH-Libr. Zurich.

1944-09-06...	German V2-rocket attacks on London and Anvers.
1944-autumn	*Work on the Hamburg betatron is continued by Kollath and Schumann.*
1944-10	*Another BBC meeting in Heidelberg to discuss the 200-MeV-betatron (according to Kaiser [Ka47]), present Dr.Meyer-Delius (Dir. BBC), Otto Weiss, Dr. Helmut Boecker plus Wideröe und Kollath representing the 'Mega-volt-Test-Laboratory' (MVA).*
1944-autumn	The Rheotron-Laboratory of Schmellenmeier is moved from Berlin to Oberoderwitz in Oberlausitz (near the Czechoslovak border).
1944-autumn	*RW participates to a meeting at the Kaiser-Wilhelm-Institut in Berlin (chairman Heisenberg), where, among other matters, the betatron is declared useless for war purposes. However, it is recommended that its development for medical applications and research in nuclear physics should continue.*
1944-11	*RW visits the betatron laboratory at 'Siemens-Reiniger-Werke' in Erlangen after which Siemens appear to have switched to 50-Hz-operation of betatrons (the first one was operated at 550 Hz).*
1944-End	Touschek in Gestapo jail in Hamburg-Fuhlsbüttel, after being discovered reading foreign magazines in the Hamburg Chamber of Commerce. He is, however, allowed to continue working. *RW and colleagues provide Touschek with his books, some food and cigarettes, but cannot get him free.* In jail Touschek develops a theory of radiative dumping for electrons circulating in betatrons [Am81].
1945-Beg.	*End of R&D for a 200-MeV-betatron at BBC, according to the Kaiser-report [Ka47] p. 8.*
1945-02-13	Allied air attack destroys Dresden.
1945-02	Werner von Braun leaves Peenemünde with 500 engineers and 14 tons of documents. They transfer south [Jo79] and hide the documents in a mine in the Harz.

1945-02(?)	Touschek to be transferred from Hamburg jail to Kiel. During the march he falls, is shot by a guard, and left for dead. He recuperates and is again imprisoned in Altona jail [Am81].
1945-02	*RW applies for three more German patents on betatrons.*
1945-02(?)	Following instructions from the German Aviation Ministry the Hamburg-betatron is transferred to Kellinghusen, near Wrist (between Bad Bramstedt and Itzehoe, 40 km north of Hamburg) [Gi93]. It works as well as it did in Hamburg.
1945-03-28	The Rheotron-Laboratory is transported by lorry to Burggrub, a small town in 'Kreis Ebermannstadt' (passing close to Dresden in flames) between Bamberg and Bayreuth in High-Franken ([Sw92] p. 122).
1945-03	*RW receives a final payment for his work from Hollnack (38.000 RM plus 38.000 NKr) and returns to Oslo by train with several stops caused by sabotage. He had his documents cleared in Copenhagen.*
1945-03	RW's brother Viggo is freed by American troops near Darmstadt.
1945-03-27	Last of 2,800 V2 fired [Jo78] [Jo79].
1945-03-29	Last of 10,500 V1 lounched [Jo78] [Jo79].
1945-04-14	US-Troops free Richard Gans and take over the Rheotron-Laboratory of Schmellenmeier in Burggrub.
1945-04-30	Hitler commits suicide in the Führerbunker.
1945-05	German troops retreat from Norway.
1945-05-03	British troops occupy Hamburg without a fight.
1945-05-07	Unconditional surrender; end of the War.
1945-05-09	Quisling surrenders to Norwegian Police.
1945-05	Hollnack makes arrangements with the British troops. Kollath, Schumann and Touschek can continue working with the 15 MeV betatron in Kellinghusen.
1945-05-23	*RW is arrested in Oslo (Ilebu jail), accused of having worked on the develoment of V2-rockets. In jail he writes a detailed report on the Hamburg betatron.*

1945-06?	Touschek is liberated from prison by the British authorities. He goes to Kellinghusen where he writes several additional theoretical reports on the betatron [To45].
1945-07-09	*G. Randers visits RW in prison to clarify his activities during the War. On the same day there is a solar eclipse over Europe.*
1945-07-09	*RW is freed, after 48 days, following an intervention by 'a friend of Odd Dahl' (G. Randers?) and probably other prominent scientists [Da81].*
1945-07...	*Until the spring of 1946 RW has no job in Oslo, no money, no passport. NEBB stops paying his salary. RW develops the theory of the gigator (the 'synchrotron').*
1945-08-06	Atomic bomb dropped on Hiroshima.
1945-09-05	McMillan presents the synchrotron-principle [Mc45].
1945	Veksler presents the synchrotron-principle [Ve45].
1945-11	*An ad hoc commission of experts to provide a professional assessment of RW is formed in Oslo.*
1945-12-11	Kollath reports on the betatron tests in Wrist [Ko45].
1945-12	Conclusion of betatron-tests in Wrist. The 15-MeV-betatron is then transported to the Woolwich Arsenal near London. Kollath helps to run it there. It is used to X-ray iron plates after which all trace of this machine vanishes, it has probably been dismantled and scrapped.
1946-01-31	*RW applies for a Norwegian patent in which the synchrotron principles are described with many details [Wi46]. Privately submitted through an agency (not for BBC), the 'Tandbergs Patentkontor' Oslo.*
1946-02-14	*An experts' report on the activity of RW during the War is presented to the Norwegian Police [Hy46]. It is evidently inspired by the overheated patriotic feelings of the time and includes assumptions (in part incorrect, due to lack of information) which are not taken into account by the Authorities [Wa94].*

1946-Beginn.	Touschek moves to Göttingen, attracted by the installation of Gund's 6 MeV betatron and starts his diploma-thesis.
1946-spring	*RW is given a provisional Norwegian passport for one month.*
1946-05-15	*RW applies for a Swiss patent on the principles of the synchrotron (253582).*
1946-?	Goward and Barnes succed testing a first synchrotron.
1946-Easter	*RW spends approximately two weeks in Baden. Professor Paul Scherrer, a friend of Theodor Boveri, recommends RW for a position at BBC-Basel. RW starts designing a betatron for 31 MeV with H. Hartmann. There is an agreement on future work on betatrons at BBC.*
1946-summer	Touschek obtains his title of 'Diplomphysiker' in Göttingen with a thesis on the theory of the betatron, supervised by R. Becker und H. C. Kopferman [Am81].
1946-08-01...	*RW starts working for BBC, Baden (CH) and receives a salary as of August 1. In need of money, he later sells the rights for the Norwegian synchrotron patent to BBC (for about 10.000 sfr) with the legal advice of Ernst Sommerfeld and Otto Lardelli (BBC).*
1946-08-19	*RW and his family move from Oslo to Zurich. By boat to Anwers with their car. They initially move into a flat in Zurich.*
1946-10	*RW is called back to Norway to take part in a judicial hearing. He stays with his parents.*
1946-11-02	*RW accepts a 'forelegg' with minor allegations about his behaviour during the War [Wa94] (it includes a fine and confiscation of most of the last instalment of money received from Hollnack). By doing so he avoids a court trial. He is immediately authorised to return to Zurich, with a passport valid only for Zurich.*
1946-11	*RW returns to Zurich and continues working on the construction of the BBC-betatron for 31 MeV electrons.*

1946-11...	*Up to 1986 a total of 78 BBC-betatrons are installed worldwide. Their main use is for medical therapy (cancer treatment), but some are used for materials tests.*
1947-01	Hermann F. Kaiser from the US Naval Research Lab. Washington DC reports on European developments on induction accelerators [Ka47]: Gund's and *Wideröe's* work is described. *One patent application from Siemens and seven from 'C. H. F. Müller, Dr.Müller' are mentioned (1942-1945). Kaiser considers RW's 200 MeV project as the most promising of the time in Europe and provides many details, including cost estimates.*
1947...	*RW and his family live in Zurich until 1948, not very comfortably, freezing... RW is working at BBC in Baden. Ragnhild complains that he works too hard.*
1947-03	*RW obtains an ordinary Norwegian passport, valid for all countries, and starts travelling. He keeps accurate notes of all trips, conferences attended, visits and meetings.*
1947-04-21	*RW submits a short comment to 'Journ. of Appl. Phys.' correcting some statements contained in the Kaiser-Report [Wi47a].*
1947-05-22	Rudolf Kollath and Gerhard Schumann submit their article describing the 15-MeV-betatron and its performance to 'Archiv f. Elektrot.' [Ko47]. It includes important information and many details.
1947-08	Gund's 5-MeV-betatron is successfully operated in Göttingen. Up to 70% of the electron beam is extracted following a 'scattering' procedure [Gu49].
1948	The Radiumspital in Oslo orders a 6-MeV-betatron from Siemens Erlangen.
1948	*RW starts the 31-MeV-betatron project for the Kantonsspital Zurich.*
1948-11-09	*RW's applies for a German patent (now BBC) on the principles of the synchrotron. It is published on Aug. 21, 1952 (847318) and gives recognition to the Norwegian patent 76696 submitted on Jan. 31, 1946.*

1949	*RW's family moves from Zurich to Baden.*
1949-autumn	*Installation of the first BBC-betatron (31 MeV) at the Kantonsspital Zurich.*
1949-autumn	Olav Netteland from Oslo's Radiumspital visits Erlangen and finds no significant progress on the 6-MeV-betatron. Siemens working on a larger one.
1950	Radiologists Congress in London.
1951-04	*Inauguration and start of operation of the first BBC 31 MeV betatron at the Kantonsspital Zurich. First patients are irradiated.*
1951-09	*Netteland and Dr. Steen from the Radiumspital in Oslo visit the Kantonsspital Zurich and see the 31-MeV-betatron in operation.*
1951-autumn	*Dr. Eker orders a betatron from BBC for the Radiumspital in Oslo.*
1952	The Cosmotron accelerator in Brookhaven reaches a particle energy of 3,000 MeV (= 3 GeV).
1952-05-05	*(to 1952-50-08) First meeting of the Council of the future CERN in Paris. A provisional CERN-PS-Group is formed to plan a 10-GeV proton-synchrotron; members are: Odd Dahl (chairman), H. Alfven, W. Gentner, F. Goward, F. Regenstreif. RW is appointed as part-time adviser (he is not present).*
1952	*A 31-MeV-betatron from BBC is installed in the Inselspital in Berne.*
1952-summer	*A 31-MeV-betatron from BBC is installed in the Radiumspital in Oslo. Six months later it is operational.*
1952-06-03	*(to 1952-06-19) International Conference in Copenhagen to discuss future projects on nuclear and particle physics for Europe. RW joins on 1952-06-17; he does not meet Odd Dahl there.*
1952-06-20	(to 1952-06-23) Second meeting of the provisional CERN-Council in Copenhagen. The PS-Group welcomes new members D. W. Fry, K. Johnsen und Chr. Schmelzer.

1952-08-04	*Returning from Australia via the USA, RW meets (for the first time) Odd Dahl in Brookhaven. Until Aug. 10, 1952 the three CERN 'delegates' (RW, Dahl and Goward) discuss with Courant, Livingston, J.Blewett and Snyder on their newly developed principle of 'strong focusing'.*
1952-10-04	(to 1952-10-07) Third meeting of CERN-Council in Amsterdam. A 30-GeV-synchrotron with modern 'strong focusing' is proposed for CERN.
1952-11-04	*RW applies for German and Swiss patents on the extraction of electrons from betatrons. German No. 954814, made public on Dec. 1956 [Wi52].*
1952-12-18	*RW, Citron und Gentner visit the future site of CERN in Meyrin, north of Geneva.*
1953-03-26	*RW's German patent on 'storage rings' (1943) is retrospectively approved and published.*
1953-12-12	*Inaugural lecture of RW at ETH in Zurich.*
1954-05-17	Start of works for CERN in Meyrin.
1954-07-15	*RW becomes head of the department 'Electric Accelerators' (EA) at BBC.*
1954-10-18	*RW to Mannheim and Karlsruhe to negociate a deal on users rights for Steenbeck's patents at a German Federal Court. BBC eventually has to pay 100,000 DM to Siemens. BBC is represented by lawyer Otto Lardelli. According to RW, the historical facts are not correctly taken into account.*
1954	In the BEVATRON accelerator in Berkeley, protons reach an energy of 6,1 GeV.
1955	Kollath publishes the first edition of his book on particle accelerators, Vieweg Publishers, Braunschweig [Ko55].
1955	*RW's family moves from Baden to Nussbaumen*
1955-06-10	Corner-stone laying for the European CERN-Laboratory in Meyrin, north of Geneva.
1956-01-23	Kerst et al. [Ke56] propose synchrotrons with strong focusing to be used as storage rings.

167

1956	At the 'CERN Symposium on High Energy Accelerators and Pion-Physics' Gerry O'Neill proposes 'The Storage-Ring-Synchrotron' [O´N56]. *RW is present and describes his ideas on storage rings in a discussion [Wi56]. RW meets O'Neill there.*
1956...	*BBC (RW) starts constructing the Turin synchrotron for 105 MeV electrons (with Gonella, Gleb Wataghin and others). It is a synchrotron with initial betatron regime.*
1956-12-20	*RW's German patent on the extraction of electrons from betatrons is accepted and published.*
1957	*Successful extraction of electrons from the betatron at the Inselspital in Berne.*
1959-1963	*RW is contracted as adviser to DESY (Synchrotron).*
1959	*The 32-MeV mobile betatron for the private clinic 'Casa di Cura S. Ambrogio' (Prof. Dr. Cova) in Milan is installed (named 'Asclepitron'). Still in operation in 1990.*
1959-11-24	The CERN-Proton-Synchrotron (28 GeV) is commissioned and starts operation.
1959-12-18	The research centre DESY in Hamburg is founded. A synchrotron for 6,4-GeV-electrons is under construction.
1960	The Brookhaven 31-GeV-synchrotron starts operation.
1960-03-07	Bruno Touschek presents his proposal for the first electron-positron storage ring (AdA) at Frascati [To60].
1961-02-27	AdA starts operation in Frascati.
1962-07-10	*RW receives a 'Dr.h.c.' from the RWTH Aachen.*
1962	*RW becomes 'Titular-Professor' at the ETH Zurich.*
1962...	*RW's main interest: The effects of radiation on living cells. He develops a 'Two-Components-Theory'.*
1962	Kollath's book on accelerators: 2nd edition [Ko62].
1964-04	*RW receives the 'Dr.med.h.c.' from Zurich University.*
1966	*RW's thesis of 1928 appears in English (translated at DESY) in a book edited by Stan Livingston [Li66].*

Year-Month-Day

1969	*RW retires from BBC - but continues working.*
1969-05-03	*RW receives the 'Röntgenmedaille' of the City of Remscheid.*
1970	BBC-betatrons for 45 MeV.
1971-01-24	*RW receives the 'Röntgenpreis' from the City of Würzburg and the Physical and Medical Society of Würzburg.*
1972	*RW's final lectures at the ETH Zurich.*
1973	*Gold Medal at the XIII JRC in Madrid.*
1973	*RW becomes a member of the Norwegian Academy of Sciences.*
1981	Odd Dahl publishes 'Trollmann og rundbrenner' (an autobiographic book) [Da81].
1982-01-10	*RW's lecture at the University of Oslo about his life and scientific work. Until Jan. 17, Conference at Geilo.*
1982-07-10	*An article by Olav Aspelund on RW is published in 'Morgenbladet' Oslo [As82].*
1983	*Finn Aaserud and Jan Vaagen publish a longer Article on RW in the Norwegian magazine »Naturen« [Aa83], after an interview in Oslo (see [Wi91]).*
1984-02	*RW's retrospective article in Europhys. News [Wi84].*
1984	*RW becomes honorary member of ESTRO.*
1992-03	*Per Dahl, son of Odd Dahl, reports on RW's life and work in an SSC-Report [Da92] (10 pages).*
1992-04	*RW is awarded the Robert-Wilson-Prize of the American Physical Society APS.*
1992-07-11	*RW celebrates his 90th birthday in Oslo.*
1992-07	*RW is honorary chairman in a session of the International Conference on High Energy Accelerators in Hamburg.*
1992-12-02	*A Symposium to celebrate RW's 90th anniversary takes place at the ETH in Zurich.*

References

[Aa83]   Aaserud, F. og Vaagen, J.: 'Et møte med Rolf Widerøe, den første akseleratordesigner', Naturen, No. 5-6, 191-196 (1983).

[As82]   Aspelund, O.: 'Rolf Wideröe – Scientist of exceptional dimensions', Morgenbladet, Oslo, Saturday, July 10, 1982.

[Am81]   Amaldi, E.: 'The Bruno Touschek Legacy', CERN-Report No. 81-19, 83 pages (1981).

[Ba60]   Barendsen, G.W.: in 'The Initial Effects of Ionising Radiation on Cell', Academic Press London (1961), p. 183.

[Be62]   Bender, M.A. and Gooch, P.C.: J. Rad. Biol. 5, 133 (1962).

[Be90]   Bergmüller, H.: 'Eine Erinnerung an C.H.F.Müller', published by Philips Medizin Systeme GmbH, Hamburg (1990).

[Bi26]   Biermanns, J.: 'Überströme in Hochspannungsanlagen', book, Springer (1926).

[Br27]   Breit, G. and Tuve, M.A.: Carnegie Institution Year Book 27, 209 (1927/28) (on betatron tests).

[Br28]   Breit, G., Dahl, O., Hafstad, L.R. und Tuve, M.A.: (Carnegie Institution), on Tesla-coil high voltage devices; Nature 121, 535 (1928), Phys. Rev.: 35, 51 (1930); 35, 66 (1930); 35, 1406 (1930); 36, 1261 (1930).

[Br30]   Brasch, A. und Lange, F.: Naturwiss., 18, 769 (1930); Z. Physik, 70, 10 (1931) (on experiments with high voltage).

[Cl87]   Close, F., Marten, M. und Sutton, Ch.: 'The Particle Explosion', Oxford University Press (1987), 239 pages.

[Ch50]   Christofilos, N.: 'Focusing System for Ions and Electrons' US-patent No. 2,736,799, submitted on March 10, 1950, issued on Febr. 28, 1956, reprinted in [Li66] on p. 270.

[Co32]   Cockroft, J.D. and Walton, E.T.S.: Proc. Roy. Soc. (London), A136, 619 (1932); A137, 229 (1932); A144, 333 (134).

[Co52]   Courant, E.D., Livingston, S.L. and Snyder, H.S.: 'The Strong Focusing Synchrotron – A New High Energy Accelerator', Phys. Rev. 88, 1190-1196 (1952), reprinted in [Li66] p.262.

[Co83]   Cohen, L.: 'Biophysical Models in Radiation Oncology', CRC-Press Inc., Boca Raton, Florida, 1983.

[Da81]   Dahl, O.: 'Trollmann og rundbrenner' Gyldendal Norsk Forlag – Oslo (1981), Autobiography, 228 pages.

[Da92]   *Dahl, P.F.*: 'Rolf Wideröe: Progenitor of Particle Accelerators', SSC-Report SSCL-SR-1186, 10 pages (1992).

[Ec93]   *Eckert, M.*: 'Die Atomphysiker – Eine Geschichte der theoretischen Physik am Beispiel der Sommerfeld-Schule', Vieweg Braunschweig (1993), 308 pages.

[Fe81]   *Fehr, W.*: 'C.H.F.Müller ...mit Röntgen begann die Zukunft', published by C.H.F.Müller, 87 pages, 1981.

[Fo93]   *Ford, J.*: 'Little Linacs Fight Cancer', Beam Line (SLAC), **23**, Nr. 1, p. 6-13 (1993).

[Gi93]   *Giel, R,*: letters to R. Wideröe and P. Waloschek (1993)

[Go46]   *Goward, F.K. and Barnes, D.E.*: Nature, **158**, 413 (1946).

[Go64]   *Gonella, L., Nabholz, H. and Wideröe, R.*: 'The Turin 100-MeV Electron Synchrotron' Nuclear Instr. & Methods, **27** 141-155 (1964).

[Gr31]   *Graaff, R. J. Van de*: Phys. Rev. **38**, 1919A (1931). First of a series of papers on electrostatic high voltage generators.

[Gr21]   *Greinacher, H.*: Z. Physik **4**, 195 (1921) (cascade generator).

[Gu46]   *Gund, K.*: Dissertation Göttingen (1947) (unpublished).

[Gu49]   *Gund, K. und Reich, H.*: Z. Physik **126**, 383 (1949).

[Gu50]   *Gund, K. und Paul, W.*: 'Experiments with a 6-MeV-Betatron', Nucleonics, **7**, 37 (July, 1950).

[Ha65]   *Haissinski, J.*: 'Expériences sur l'anneau de collisions AdA', Thèse, Orsay Serie A, No. 81, soutenue le 5 février 1965.

[He87]   *Hermann, A., Krige, J., Mersits, U. and Pestre, D.*: 'History of CERN', North Holland Amsterdam Vol. 1 and 2 (1987).

[Hy46]   *Hylleraas, E.A., Randers, G., Tangen, R. and Wergeland, H.*: A report on Wideröe's activity during the War written for the Oslo Police. (text in Norwegian), 15 pages, typed; Febr. 14, 1946; Norw. National Archives Oslo (see [Wa94]).

[Is24]   *Ising, G.*: 'Prinzip einer Methode zur Herstellung von Kanalstrahlen hoher Voltzahl' (in German), Arkiv för matematik o. fysik, **18**, Nr. 30, 1-4 (1924).

[Is33]   *Ising, G.*: 'Högspänningsmetoder för atomsprängning' in Kosmos, Annuary of the Swedish Phys. Assoc. (1933).

[Ja36]   *Jassinski, W.W.*: Arch. f. Elektrot., **30**, 590 (1936).

[Ja84]   *Jacob, M. and Johnsen, K.*: 'A Review of Accelerator and Particle Physics at the CERN Intersecting Storage Rings', CERN 84-13, Nov. 30, 1984.

[Jo78]   *Jones, R.V.*: 'Most Secret War', Hamish Hamilton Ltd., 1978, Coronet Edition 1979, 702 pages.

171

[Jo79]   *Johnson, B.*: 'The Secret War' British Broadcasting Co. (BBC) book 1978, paperback 1979, 352 pages.

[Ka47]   *Kaiser, H.F. (U.S. Naval Research Lab., Washington, D.C.)*: 'European Electron Induction accelerators', J. of Appl. Phys. **18**, 1-17 (1947). Review, with many details of the work of Gund and Wideröe and on the BBC 200-MeV betatron project.

[Ke40a] *Kerst, D.W.*: 'Acceleration of Electrons by Magnetic Induction', Letter to the Editor of Oct. 15, 1940, Phys. Rev., **58**, 841 (1940) and 'Induction Electron Accelerator', Comm. to the APS, Nov. 22-23, 1940, Phys. Rev. **59**, 110 (1941).

[Ke40b] *Kerst, D.W. (General Electric)*: 'Magnetic Induction accelerator' US-patent No. 2,297,305, submitted on Nov. 13, 1940, issued on Sept. 29, 1942.

[Ke41a] *Kerst, D.W.*: 'The Acceleration of Electrons by Magnetic Induction', Phys.Rev. **60**, 47-53 (1941), reprinted in [Li66].

[Ke41b] *Kerst, D.W. and Serber, R.*: 'Electronic Orbits in the Induction accelerator', Phys.Rev. **60**, 53 (1941), reprinted in [Li66].

[Ke42]   *Kerst, D.W.*: '20 MeV Betatron or Induction accelerator', Jour. Sci. Instr. **13**, 387-394 (1942).

[Ke46]   *Kerst, D.W.*: (Historic Review) Nature **157**, 90 (1946).

[Ke56]   *Kerst, D.W., Cole, F.T., Crane, H.R., Jones, L.W., Laslett, L.J., Ohkawa, T., Sessler, A.M., Symon, K.R., Terwilliger, K.W. and Nilsen, N.V.*: 'Attainment of Very High Energy by Means of Intersecting Beams of Particles', (subm. on Jan. 23, 1956), Phys. Rev. (Letters) **102**, 590-591 (1956).

[Ko45]   *Kollath, R.*: Report 11.12.45, ETH-Libr. Hs 903: 29 (5 pages).

[Ko47]   *Kollath, R. und Schumann, G.*: 'Untersuchungen an einem 15-MeV-Betatron', Z. Naturforschg. **2a**, 634-642 (1947).

[Ko55]   *Kollath, R.*: 'Teilchenbeschleuniger', Vieweg Verlag Braunschweig, 1st ed. 1955 (222 pages), 2nd improved ed. with several co-authors 1962 (335 pages).

[La30]   *Lawrence, E.O. and Edlefsen, N.E.*: Science, **72**, 376 (1930)

[La31a] *Lawrence, E.O. and Sloan, D.*: Proc. Nat. Ac. Sc., **17**, 64 (1931) and 'The Production of Heavy High Speed Ions without the Use of High Voltages', Phys. Rev. **38**, 2022 (1931) reprinted in [Li66], p. 151.

[La31b] *Lawrence, E.O. and Livingstone M.S.*: (first cyclotron in operation; 80 keV, 13 cm) Commun. to the Am. Phys. Soc., in May, 1931; Phys. Rev **37**, 1707 (1931).

[Le18] *Lenard, Ph.*: 'Quantitatives über Kathodenstrahlen aller Geschwindigkeiten', in 'Abhandlungen der Akademie der Wissenschaften', Heidelberg 1918 (2. ed. 1925); Carl Winter Universitätsbuchhandlung, p. 1 to 258 + tables.

[Li31] *Livingston, M.S.*: 'The Production of high-velocity Hydrogen Ions without the Use of High Voltages', PhD thesis, University of California, April 14, 1931.

[Li62] *Livingston, M.S. and Blewett, J.P.*: 'Particle Accelerators', McGraw-Hill Book Company, Inc. (1962), 666 p.

[Li66] *Livingston, M.S.*: 'The Development of High-Energy-Accelerators', commented reprints or translations of original papers (book), Dover Publish. Inc. N.Y. (1966).

[Mc45] *McMillan, E.M.*: 'The Synchrotron – A Proposed High Energy Particle Accelerator', Phys. Rev. **68**, 143-144 (1945) (Issue Nr. 5 and 6, Sept. 1 and 15 1945), submitted on Sept. 5, 1945; reprinted in [Li66].

[Me44] *Meyer-Delius*: BBC meeting minutes. ETH-Libr. **Hs 903**: 63

[Mu43] *'Dr. Müller'*: 'Über die Elektronenerzeugung im Strahlentransformator', typed report, drawings, ET-Libr. **Hs 903:** 47.

[O'N56] *O'Neill, G.*: 'The Storage Ring Synchrotron', Proc. Int. Conf. on High Energy Accelerators, Vol.1, p. 64-67 (1956).

[O'N59] *O'Neill, G.*: 'Storage Rings for Electrons and Protons', Proc. Int. Conf. on High-Energy Accelerators and Instrumentation', CERN, Geneva 1959, pages 125-136.

[O'N76] *O'Neill, G.*: 'The High Frontier, Human Colonies in Space' (book), William Morrow New York (1976), 283 pages; ref. on Wideröe on p. 239.

[Os87] *Osietzki, M.*: 'Das Liliput-Zyklotron – ein vergessenes Projekt', Kultur und Technik, **3**, 182-187 (1987).

[Os88] *Osietzki, M.*: 'Kernphysikalische Großgeräte... Teilchenbeschleuniger bei Siemens 1935-45' in 'Technikgeschichte', **55**, p. 25-46 (1988).

[Pa47] *Paul, W. und Dänzer, H.*: 'Betatrons', Chap. 5.5 in 'Naturforschung und Medizin in Deutschland 1939-1946' Vol. **14**, Part II, p. 49-80, Chemie-Publishers, Weinheim (1947).

[Pa79] *Paul, W.*: 'Early Days in the Development of Accelerators', Proc. Internat. Symposium in Honor of Robert R. Wilson, April 27, 1979.

[Pa93] *Paul, W.*: letter to P. Waloschek from Jan. 14, 1993.

[Pu82] Purohit, S.C. and Pohlit, W.: Int. Journ. Radiation and Onkol. Biol. Phys. **8**, 495-499 (1982).

[Ru29] Rüdenberg, R. *(Herausgeber)*: 'Relais und Schutzschaltungen in elektrischen Kraftwerken und Netzen' (book) Julius Springer Berlin 1929, 284 pages.

[Ru33] Rüdenberg, R. und Steenbeck, M. *(Siemens)*: German-Pat. No. 656,378, submitted on March 1, 1933, issued Febr. 4, 1938.

[Sa93] Sand, Helene: on Viggo Widerøe, Illustrated Weekly 'Perspektiv', Oslo, August 15, 1993, No. 4, p. 32-33.

[Sch72] Schumacher, W.: 'Nutzbarmachung neuer Erkenntnisse über die Fraktionierung bei der Bestrahlung bösartiger Tumoren für die Praxis' Röntgen-Berichte, Vol *1*, p. 91-100 (1972)

[Se58] Sempert, M.: 'Das Brown Boveri 31-MeV-Betatron für zerstörungsfreie Werkstoffprüfung', Brown Boveri Mitteilungen, **45**, 383-396 (1958).

[Se80] Segrè, E.: 'From X-Rays to Quarks – Modern Physicists and their Discoveries' (book) San Francisco, Freeman 1980, 337 p.

[Sl22] Slepian, J.: 'X-Ray Tube' US-Pat. 1,645,304, submitted on April 1, 1922, published on Oct. 11, 1927. Submitted in Germany too, published 1928.

[St35] Steenbeck, M. *(Siemens)*: German patent No. 698,867, submitted on March 7, 1935, published on Dec. 6, 1940 (Austrian Patent No. 153,324).

[St36] Steenbeck, M. *(Siemens)*: US-patent 2,103,303, submitted on March 3, 1936, published on Dec. 28, 1937.

[St42] Steenbeck, M.: Article in 'Electronics', February 1942, p. 22-23.

[St43] Steenbeck, M.: 'Beschleunigung von Elektronen durch elektrische Wirbelfelder', Naturwissenschaften, Issue 19/20, **31**, 234 (1943).

[St67] Steenbeck, M.: 'Impulse und Wirkungen – Schritte auf meinem Lebensweg', (book) Verlag der Nation Berlin DDR (1977), 447 pages.

[Sw92] Swinne, E.: 'Richard Gans – Hochschullehrer in Deutschland und Argentinien', (book) Berliner Beiträge zur Geschichte der Naturwiss. und der Technik, ERS-Verlag, Berlin (1992).

[Sw93] Swinne, E.: letters to P. Waloschek (1993).

[Ta93] Tangen, R.: letters to P. Waloschek (1993).

[To45] Touschek, B.: typed reports on the betatron (in German), ETH-Libr. **Hs 903**: 29, 30, 33, 73, 74.

[To60] *Touschek, B.*: First proposal to built an electron-positron storage ring, presented at Frascati, March 7, 1960 [Am81].

[Ve45] *Veksler, V.*: 'A New Method of Acceleration of Relativistic Particles', Journal of Physics UdSSR, **9**, 153-158 (1945), English translation in [Li66].

[Wa29] *Walton, E.T.S.*: Proc. Cambr. Phil. Soc. **25**, 469-481 (1929).

[Wa91] *Waloschek, P.*: 'Reise ins Innerste der Materie – Mit HERA an die Grenzen des Wissens', 280 pages, DVA Stuttgart (1991).

[Wa93] *Waloschek, P.*: 'Wideröe über Wideröe – Ein Zeitzeuge berichtet', Video, 55 min., from October 1992, in German. VHS copies can be obtained from Nick Wall Photography, Achter Lüttmoor 45, D - 22559 Hamburg.

[Wa94] *Waloschek, P.*: 'Some remarks on Wideröe's Biography', DESY Internal Report, 1994 (in preparation).

[We45] *Westendorp, W.F.*: J. Appl. Phys. **16**, 657 (1945).

[Wi23] *Wideröe, R.*: original copy-books from 1923 to 1928, with sketches and computations on the ray-transformer, ETH-Libr. Zurich, **Hs 903**: 633-638.

[Wi24] *Wideröe, R.*: 'Inflationsanalyse', report on the inflation 1922-23, Statsökonomisk Tidskrift (Oslo 1924) p.189-206.

[Wi28] *Wideröe, R.*: 'Über ein neues Prinzip zur Herstellung hoher Spannungen', Arch. f. Elektrot. **21**, 387 (1928) (Wideröe's dissertation in Aachen); English in [Li66].

[Wi37] *Wideröe, R.*: 'Über technische Probleme der gekuppelten Kraftwerke in Ostnorwegen', in 'E und M Wien' (Elektrotechnik und Maschinenbau), **55**, 617-624 (1937).

[Wi42] *Wideröe, R.*: 'Der Strahlentransformator', Arch. f. Elektrot., **37**, 542-555 (1943), submitted on Sept. 15, 1942.

[Wi43a] *Wideröe, R. (BBC)*: German patent No. 876,279 (on storage rings); submitted on Sept. 8, 1943, issued on May 11, 1953.

[Wi43b] *Wideröe, R.*: Report on the historic development of betatrons, many references, from Oct. 1, 1943, ETH-Libr. **Hs 903**: 46.

[Wi43c] *Wideröe, R.*: Report 'Fast building of betatrons in Germany', from Oslo, Nov. 6, 1943, 4 pages, ETH-Libr. **Hs 903: 48**.

[Wi44] *Wideröe, R.*: Report on visits at BBC Weinheim from April 27 to 29, 1944, dated May 1, 1944; ETH-Libr. **Hs 903:** 62 and 63.

[Wi46] *Wideröe, R.*: 'Anordnung zur Beschleunigung von elektrisch geladenen Teilchen' (theory of the synchrotron), Norwegian Patent (No. 76,696) submitted on Jan. 31, 1946, in Germany on Nov. 9, 1948 (No. 847,318), published on June 26, 1952.

[Wi47a] *Wideröe, R.*: 'European Induction accelerators', Journ. of
Appl. Physics, **18**, 783 (1947), some remarks on [Ka47].

[Wi47b] *Wideröe, R.*: 'The Gigator – a Proposed New Circular Accelerator for Heavy Particles', Phys. Rev. **72**, 978 (1947).

[Wi49] *Wideröe, R. (BBC)*: 'Einrichtung zur Beschleunigung von
elektrisch geladenen Teilchen' (betatron) German-patent
No. 856 491, published Nov. 20, 1952; submitted first in
Switzerland on Sept. 25, 1949.

[Wi52] *Wideröe, R.*: German patent on beam extraction, Nr. 954,814;
submitted Nov. 4, 1952, issued Dec. 20, 1956.

[Wi53] *Wideröe, R.*: 'Das Betatron', Z. f. angewandte Physik, **5**,
187-200 (1953) (with many references).

[Wi56] *Wideröe, R.*: Discussion, Proc. Int. Conf. on High Energy
Accelerators, Geneva, Vol.*1*, p. 97-98 (1956).

[Wi59] *Wideröe, R.*: 'Physik und Technik der Megavoltbestrahlung',
in 'Strahlenbiologie - Ergebnisse 1952-1958' edited by
H. R. Schinz:, Thieme Verl. 1959, p. 289-360, 92 refs.

[Wi62] *Wideröe, R.*: 'Grundlage und Technik der Megavolttherapie',
Ärztliche Forschung, **16**, I/598-I/614 (1962).

[Wi64] *Wideröe, R.*: 'Die ersten zehn Jahre der Mehrfachbeschleu-
nigung', (talk given in Jena) Wissensch. Zeitschr. der
Friedrich-Schiller-Universität Jena, **13**, 431-436 (1964).

[Wi70] *Wideröe, R.*: Collection of Publications, Manuscripts and
Patents, 17 Volumes, History of Science Collections, ETH-
Libr. Zurich: No. R 1970/461: 1-17 (**Hs**). Many additional
documents are conserved in the same collection.

[Wi84] *Wideröe, R.*: 'Some Memories and Dreams from the Childhood of Particle Accelerators', Europhysics News, **15**, 9-11
(1984).

[Wi90] *Wideröe, R.*: 'Zweikomponententheorie und Strahlentherapie',
Lecture at Paul-Scherrer-Inst., Villigen, 1988, published in
Strahlenther. Onkoll **166**, 311-316 (1990).

[Wi91] *Wideröe, R.*: Manuscripts; interview with Aaserud and Vaagen
(Oslo) on July 11, 1983 (76 pages), plus their article [Aa83]
(20 pages); German and Norwegian versions at the ETH-Libr.
Zurich; 62 letters to P. Waloschek (1990-1994).

Note: Some of the references were used to compile this report but are
not explicitly quoted in the text.

Appendix

Wideröe's dissertation [Wi28], including the results of the first operational drift-tube as well as the proposal for a betatron, was published in a well known periodical and later in an English translation. It therefore reached a correspondingly wide readership. The important ideas which Wideröe later submitted as patents are not as well known in research circles. This is quite natural, since patents do not in general contain scientific results but inventions; that is, ideas or technical developments for which the inventor, and usually the company employing the inventor, wish to protect their mental property by right of law.

The patenting offices check the ideas submitted for protection. These must contain some substantial technical improvement on the past, must not contradict current knowledge and must not have been previously published elsewhere. Although realisation of the idea should appear plausible, it does not require proof; the scientific value of the idea is not assessed. Only the inventor (or the company named in the patent) is permitted to use the patented ideas industrially. However, he may award or sell licences for use of the patented idea. These rights only apply for as long as the patent is valid and the required fees have been paid to the patenting office. Twenty years is the longest an idea can remain subject to patenting rights in Germany and, in general, patents are declared valid from the day of submission.

Quite different customs pertain in the field of fundamental research; scientific results and proposals are published precisely because people want them to be used or further developed by others. However, publication must have been agreed to by experts in the respective fields. The use for the scientists themselves consists in the priority which they secure by making a publication – this in turn strengthens their positions as researchers. Scientists are usually quite happy to pass on technical details because they can rarely be turned to economic advantage. New ideas are referred to as 'proposals' and not as inventions. Although patents are taken into consideration they are only rarely deemed to be works of scientific merit.

Researchers like Wideröe who work in industry often find that they must submit patents in order to maintain the legal protection required by their companies (or themselves). Usually publication is not in the company's interest. Accordingly, some of Wideröe's patents are of a rather special kind; they contain ideas for constructing accelerators which, had they appeared in scientific journals at the right time, would certainly have stimulated a great deal of interest. A facsimile of the two probably most important patents is reproduced in the following pages. The patent on pages 179 to 182 contains the first known proposal for the construction of a storage ring. Wideröe called this a 'reaction tube' or 'nuclear mill'. At that time the only type of ring accelerator available for this purpose was the betatron, the only accelerator in which particles could be kept stable on fixed orbits. The synchrotron did not yet exist. Wideröe was considering relatively small rings and very low particle energies. This is why he proposed to force particles of equal electrical charges (atomic nuclei) onto opposing orbits using electrical fields – which have a relatively weak effect on charged particles. This type of storage ring was never built. The energy would have been too low to induce nuclear reactions.

However, the text of this patent includes (without claim) a proposal whereby positive and negative charged particles would be made to turn in opposing direction with the help of magnetic fields – which have a much stronger effect. Wideröe mentions atomic nuclei (and particularly protons) as positive particles to be made to collide with negative electrons, both particle types being stored in the same ring. Although this is feasible it is not easy, and is exactly the type of installation which H. Gerke, H. Wiedemann, B. Wiik and G. Wolf proposed for DESY in 1972; protons and electrons would be stored in a single ring (DORIS) and made to collide.

However, Touschek had already realised Wideröe's idea in 1960 in Frascati, using electrons against positrons (instead of protons) and had thus put in motion the triumphant progress of this type of machine.

The second patent, which is reproduced on pages 183 to 192, contains a theory and practical ideas for the construction of synchrotrons (Wideröe called them 'gigators'). It includes many suggestions which are now regarded as the ground rules for building synchrotrons and storage rings. The number of new ideas Wideröe developed in 1945 while he was unemployed in Oslo and had time to do so, is quite astonishing (BBC bought the patent after they had employed him again in 1946). McMillan [Mc45] and Veksler's [Ve45] ideas, which they developed almost simultaneously, contain similar principles, but fewer practical suggestions.

The two patents reproduced here and a number of others can, to a great extent, be regarded as scientific contributions. They have not earned BBC much as licensable patents. Also, when larger and even industrially useful storage rings were finally being built, these patents had long lapsed. However, they are interesting documents from an historical point of view, which clearly demonstrate the astonishing level of Wideröe's thinking at that time.

P.W.

Erteilt auf Grund des Ersten Überleitungsgesetzes vom 8. Juli 1949
(WiGBL S. 175)

BUNDESREPUBLIK DEUTSCHLAND

AUSGEGEBEN AM
11. MAI 1953

DEUTSCHES PATENTAMT

PATENTSCHRIFT

Nr. 876 279

KLASSE 21g GRUPPE 36

W 687 VIIIc / 21g

Dr.-Ing. Rolf Wideröe, Oslo
ist als Erfinder genannt worden

Aktiengesellschaft Brown, Boveri & Cie, Baden (Schweiz)

Anordnung zur Herbeiführung von Kernreaktionen

Patentiert im Gebiet der Bundesrepublik Deutschland vom 6. September 1943 an
Patentanmeldung bekanntgemacht am 18. September 1952
Patenterteilung bekanntgemacht am 26. März 1953

Kernreaktionen können dadurch herbeigeführt werden, daß geladene Teilchen von hoher Geschwindigkeit und Energie, in Elektronenvolt gemessen, auf die zu untersuchenden Kerne geschossen werden. Wenn die geladenen Teilchen in einen gewissen Mindestabstand von den Kernen gelangen, werden die Kernreaktionen eingeleitet. Da aber neben den zu untersuchenden Kernen noch die gesamten Elektronen der Atomhülle vorhanden sind und auch der Wirkungsquerschnitt des Kernes sehr klein ist, wird der größte Teil der geladenen Teilchen von den Hüllenelektronen abgebremst, während nur ein kleiner Teil die gewünschten Kernreaktionen herbeiführt.

Erfindungsgemäß wird der Wirkungsgrad der Kernreaktionen dadurch wesentlich erhöht, daß die Reaktion in einem Vakuumgefäß (Reaktionsröhre) durchgeführt wird, in welchem die geladenen Teilchen hoher Geschwindigkeit gegen einen Strahl von den zu untersuchenden und sich entgegengesetzt bewegenden Kernen auf einer sehr langen Strecke laufen müssen. Dies kann in der Weise durchgeführt werden, daß die geladenen Teilchen zum mehrmaligen Umlauf in einer Kreisröhre gezwungen werden, wobei die zu untersuchenden Kerne auf derselben Kreisbahn, aber in entgegengesetzter Richtung umlaufen. Da die geladenen Teilchen dabei nicht von bei der Reaktion unwirksamen Elektronen abgebremst werden und andererseits auf einer sehr langen Wegstrecke gegen die Kerne sich bewegen können, wird die Wahrscheinlichkeit für das Eintreten der Kernreaktionen wesentlich größer und der Wirkungsgrad der Reaktion sehr stark erhöht.

Um die bei der Kreisbewegung entstehenden Zentrifugalkräfte aufzuheben, müssen die umlaufenden Teilchen von nach innen gerichteten Ablenkkräften gesteuert werden, während eine Diffusion der Teile mittels stabilisierender, von allen Seiten auf den Bahnkreis gerichteter Kräfte verhindert wird. Falls die gegeneinander umlaufenden Teilchen verschiedene Ladung haben, müssen die Ablenkkräfte mittels senkrecht zum Bahnkreis gerichteter magnetischer Felder hergestellt werden.

Sind dagegen die Teilchen von der gleichen Ladung, so muß die Ablenkung mittels eines in der Ebene des Bahnkreises radial wirkenden elektrostatischen Feldes hervorgerufen werden.

Erfindungsgemäß wird ein besonders ausgebildeter Strahlentransformator verwendet, um schnell bewegte Elektronen zur Reaktion mit Kernen zu bringen. Die Kernstrahlen, z. B. Protonenstrahlen, werden dann während der Beschleunigungszeit der Elektronen in die Kreisröhre hineingeführt.

Die für die Kernreaktion maßgebende Elektronenspannung (= Elektronengeschwindigkeit) kann dadurch gewählt werden, daß man die Kernstrahlen zu einem früheren oder späteren Zeitpunkt während der Beschleunigungsperiode in die Kreisröhre einführt. Dadurch, daß die Kerne den Elektronen entgegenlaufen, wird die relative Geschwindigkeit und somit auch die der Geschwindigkeit entsprechende Spannung noch erhöht. Um den Eintritt der Kernstrahlen in die Kreisröhre zu ermöglichen, muß dieselbe zwei entgegengesetzt gerichtete Eintrittsöffnungen besitzen.

Die Einwirkungen von schnell bewegten Protonen auf Deuteronen haben sich als besonders geeignet für die Herbeiführung von Kernreaktionen erwiesen.

Um derartige Kernreaktionen herbeizuführen, muß die Reaktionsröhre, wie bereits beschrieben, mit einer elektrostatischen Steuerung versehen werden. Das elektrostatische Feld ist dabei radial nach innen gerichtet. Um stabilisierende Kräfte mit sowohl radialen als auch axialen, d. h. senkrecht zu der Radialrichtung gerichteten Komponenten zu erhalten, soll dabei erfindungsgemäß die elektrische Feldstärke nach innen zunehmen. Man erhält hierdurch, wie Abb. 1 es zeigt, eine axiale Stabilisierungskraft. Mit 10 ist die Achse des Bahnkreises bezeichnet und mit 11 der Querschnitt durch die Kreisringröhre. Zwischen die Platten 12 und 13 ist eine Spannung des durch Plus- und Minuszeichen angegebenen Vorzeichens zu legen. Damit man auch eine radial gerichtete Stabilisierungskraft erhält, muß die Zunahme der Feldstärke langsamer als umgekehrt proportional mit dem Radius zunehmen. Man kann beispielsweise die Feldstärke proportional $1/\sqrt{r}$ sich ändern lassen, wenn r den Bahnkreisradius bedeutet. Bevor die Protonen bzw. Deuteronen in die Reaktionsröhre eingeführt werden, müssen sie eine hohe Geschwindigkeit erreicht haben.

Diese hohen Geschwindigkeiten bzw. Spannungen, in Elektronenvolt gemessen, können den geladenen Teilchen erfindungsgemäß in einem oder mehreren Strahlentransformatoren erteilt werden.

Die Beschleunigung von Protonen und Deuteronen sowie auch anderer Kerne in einem Strahlentransformator bereitet keine prinzipiellen Schwierigkeiten. Bei der Wahl der Anfangsspannung und der Vormagnetisierung des Steuerfeldes des Transformators muß man in die für Strahlentransformatoren allgemein gültige Beziehung (1); in der B, die Steuerfeldstärke, B die Induktionsfeldstärke, c die Lichtgeschwindigkeit, r wieder den Bahnkreisradius, U_0 die Anfangs-

geschwindigkeit, ε die sogenannte spezifische Massenenergie $= \dfrac{mc^2}{e}$ ($m =$ Masse, $e =$ Ladung des betreffenden Teilchens) bedeutet und den Anfangswert B_{i0} der induzierenden Feldstärke zu $-B_{im}$ angenommen ist, was besagt, daß die induzierende Feldstärke zu Anfang der Beschleunigung einen negativen Maximalwert besitzt:

$$B_i = \frac{1}{2}B_i + \left(\frac{1}{cr}\sqrt{U_0^2 + 2U_0\varepsilon} + \frac{1}{2}B_{im}\right), \quad (1)$$

für ε den für Protonen bzw. Deuteronen gültigen Wert (933 MV für Protonen und 1866 MV für Deuteronen) einsetzen. Aus der Beziehung (2) ersieht man, daß es zweckmäßig sein wird, eine möglichst hohe Anfangsspannung U_0 zu verwenden, um eine hohe maximale Spannung zu erhalten:

$$U_{max} = U_0 + \varepsilon\left(\sqrt{\left(\frac{crB_{im}}{\varepsilon}\right)^2 + 1} - 1\right) \quad (2)$$

Aus diesem Grunde wird es sich als zweckmäßig erweisen, die Protonen abwechselnd in zwei Strahlentransformatoren oder mittels einer bereits an anderer Stelle vorgeschlagenen Kaskadenschaltung von mehreren Strahlentransformatoren zu beschleunigen.

Falls man einen oder mehrere Strahlentransformatoren zur Beschleunigung der Kerne zu bringen verwendet, kann man gemäß der Erfindung zunächst die eine Art von Teilchen in dem Transformator beschleunigen und dann in die Reaktionsröhre leiten. In der nächsten Beschleunigungsperiode kann man dann die Teilchen der anderen Art, wobei die reagierenden Teilchen auch gleicher Art sein können, in demselben Strahlentransformator beschleunigen.

Die Teilchen müssen dann auf verschiedenen Wegen in die Reaktionsröhre geleitet werden, so daß sich hier eine entgegengesetzte Umlaufsrichtung ergibt (Abb. 2). Die Teilchen können auch bereits in der Kreisröhre des Transformators verschiedene Umlaufsrichtungen haben (Abb. 3). Man kann schließlich auch die beiden Teilchenarten gleichzeitig in zwei Transformatorröhren beschleunigen, die durch den gleichen magnetischen Induktionsfluß erregt werden (Abb. 4). In Abb. 2 bis 4 sind die mit I und II bezeichneten Pfeile die Fortbewegungsrichtungen der beiden Teilchenarten.

In der Reaktionsröhre rotieren bei allen gezeichneten Ausführungsformen die an der Reaktion teilnehmenden Teilchen mit konstanter Geschwindigkeit. Diese Reaktionsröhre wird demnach nicht wie die Kreisringröhre des Strahlentransformators von einem sich zeitlich ändernden Fluß durchsetzt, sondern besitzt nur ein die Fliehkraft der Teilchen aufhebendes elektrostatisches oder magnetisches Steuerfeld und ein von allen Seiten auf den Bahnkreis hin gerichtetes, eine Diffusion der Teilchen verhinderndes Kraftfeld.

PATENTANSPRÜCHE:

1. Anordnung zur Herbeiführung von Kernreaktionen, gekennzeichnet durch die Benutzung einer luftleeren Reaktionsröhre, in welcher geladene Teilchen gleicher oder verschiedener Art zum gleich-

zeitigen Umlauf in verschiedener Richtung gebracht werden.

2. Anordnung nach Anspruch 1, dadurch gekennzeichnet, daß ein Strahlentransformator während der Transformationsperioden sowohl mit Elektronen als auch mittels positiver Teilchen aufgeladen wird.

3. Anordnung nach Anspruch 2, dadurch gekennzeichnet, daß die positiven Teilchen früher oder später als die Elektronen in die Röhre eingeführt werden.

4. Anordnung nach Anspruch 1, dadurch gekennzeichnet, daß die Reaktionsröhre bei Reaktionen zwischen Teilchen gleicher Ladung elektrostatisch gesteuert wird.

5. Anordnung nach Anspruch 4, dadurch gekennzeichnet, daß das elektrostatische Steuerfeld radial nach außen abnimmt, jedoch langsamer als umgekehrt proportional mit dem Radius.

6. Anordnung nach Anspruch 1, dadurch gekennzeichnet, daß die Teilchen in einem oder mehreren Transformatoren beschleunigt werden, bevor sie in die Reaktionsröhre eingeführt werden.

7. Anordnung nach Anspruch 6, dadurch gekennzeichnet, daß die Teilchen in die Reaktionsröhre abwechselnd auf zwei verschiedenen Wegen eingeleitet werden und in dieser Weise verschiedene Umlaufsrichtungen erhalten.

8. Anordnung nach Anspruch 6, dadurch gekennzeichnet, daß die Teilchen in der Transformatorröhre verschiedene Umlaufsrichtungen haben und auch in der Reaktionsröhre verschiedene Umlaufsrichtungen besitzen.

9. Anordnung nach Anspruch 6, dadurch gekennzeichnet, daß die Teilchen in zwei verschiedenen Transformatorröhren gleichzeitig beschleunigt werden und beide Transformatorröhren von demselben Induktionsfluß erregt werden.

Hierzu 1 Blatt Zeichnungen

⊕ 5094 4.53

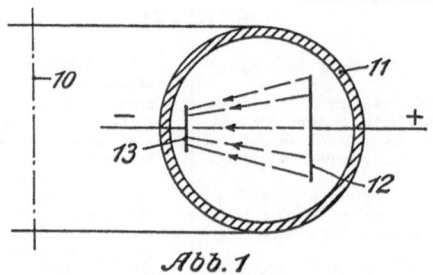

10
11
−
+
13
12

Abb. 1

I II
II
II
Transformatorröhre
Reaktionsröhre
I
I

Abb. 2

I
I
I
II
Transformatorröhre
II
Reaktionsröhre
II

Abb. 3

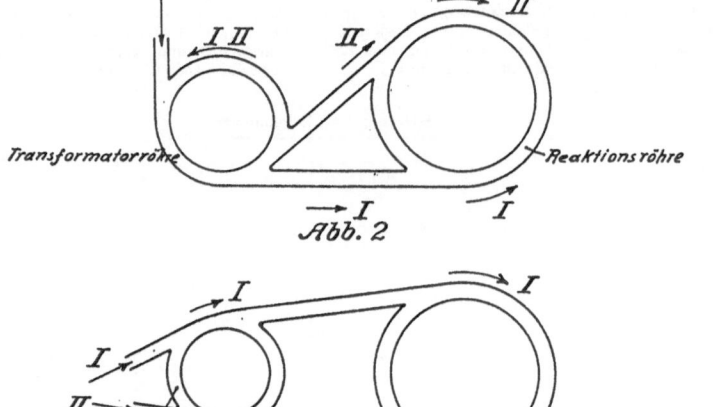

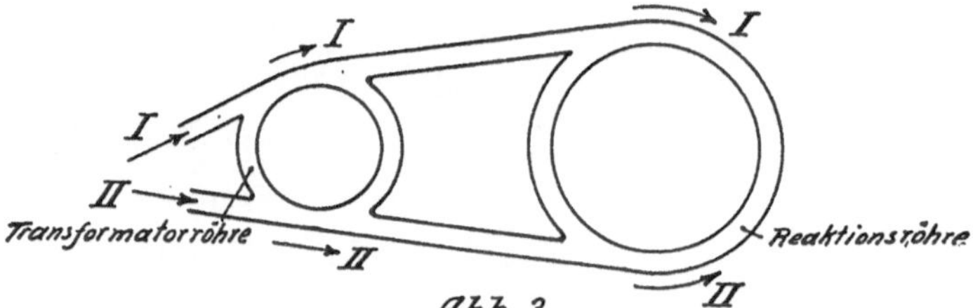

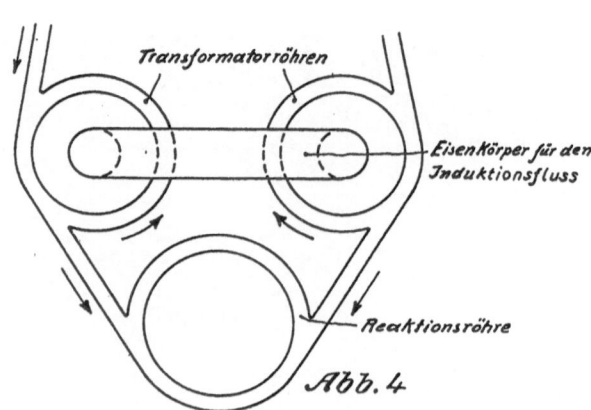

Transformatorröhren
Eisenkörper für den Induktionsfluss
Reaktionsröhre

Abb. 4

Erteilt auf Grund des Ersten Überleitungsgesetzes vom 8. Juli 1949
(WiGBl. S. 175)

BUNDESREPUBLIK DEUTSCHLAND

AUSGEGEBEN AM
21. AUGUST 1952

DEUTSCHES PATENTAMT

PATENTSCHRIFT

№ 847 318

KLASSE 21 g GRUPPE 36

$p\ 21107\ VIIIc/21gD$

Dr.-Ing. Rolf Wideröe, Ennetbaden (Schweiz)
ist als Erfinder genannt worden

Aktiengesellschaft Brown, Boveri & Cie, Baden (Schweiz)

Anordnung zur Beschleunigung von elektrisch geladenen Teilchen

Patentiert im Gebiet der Bundesrepublik Deutschland vom 9. November 1948 an
Patentanmeldung bekanntgemacht am 9.. August 1951
Patenterteilung bekanntgemacht am 26. Juni 1952
Die Priorität der Anmeldung in Norwegen vom 31. Januar 1946 ist in Anspruch genommen

---

Wenn Elektronen oder Ionen große Geschwindigkeiten erteilt werden sollen, kann dies durch Beschleunigung in Potentialfeldern geschehen, die durch eine hochfrequente Wechselspannung erzeugt werden. Dieser Vorgang ist in Fig. 1 der Zeichnung dargestellt. Eine Reihe von Zylindern ist abwechselnd an die zwei Pole für die hochfrequente Wechselspannung u angeschlossen, und die Elektronen bzw. die Ionen, die durch die Zylinder geleitet werden, werden im Raum zwischen zwei Zylindern von der Wechselspannung beschleunigt, wobei U_0 die Anfangsspannung der geladenen Teilchen ist. Durch die Wahl so langer Zylinder, daß ihre Polarität, während die Teilchen mit konstanter Geschwindigkeit durch sie hindurchgehen, wechselt, werden die Teilchen zwischen je zwei Zylindern beschleunigt und erreichen somit eine ständig höhere Geschwindigkeit, d. h. sukzessive die

kinetischen Spannungen $U = U_0 + u,\ U_0 + 2u,\ U_0 + 3u$ usw.

Diese bekannte Anordnung hat den Nachteil, daß die Zylinder infolge der hohen Geschwindigkeiten der geladenen Partikel verhältnismäßig lang werden und die Frequenz der Wechselspannung sehr hoch sein muß, damit die obengenannte Resonanzbedingung erfüllt wird. Wenn man darum hohe kinetische Spannungen erreichen will, wird diese Anordnung sehr hohe Ladeströme (Blindleistung) erfordern und wegen der Verluste entsprechend große Hochfrequenzgeneratoren. Dies schränkt das Anwendungsgebiet auf verhältnismäßig schwere Ionen, z. B. Quecksilberionen, und die Spannungen auf einige wenige MV ein.

Vorliegende Erfindung bezweckt, diesem Nachteil abzuhelfen. Sie betrifft eine Anordnung zur Beschleunigung von elektrisch geladenen Teilchen mit

183

Hilfe von in der Bewegungsrichtung der Teilchen aufeinanderfolgenden hochfrequenten elektrischen Potentialfeldern, und ist dadurch gekennzeichnet, daß die Elektroden, zwischen denen die Potentialfelder erzeugt werden, hohle Stücke der beiden Leiter einer Hochfrequenzenergieleitung sind, auf der stehende Spannungswellen erzeugt werden, wobei die geladenen Teilchen, nachdem sie ein Potentialfeld durchlaufen haben, sich innerhalb eines der genannten hohlen Stücke bewegen.

In Fig. 2 ist eine solche Anordnung gezeigt. Die beiden Leiter einer Lecherleitung werden von Zylindern a_1, b_2, a_3, b_4, a_5 und den Drahtleiterstücken b_1, a_2, b_3, a_4, b_5, die zu je einem der genannten Zylinder außerhalb des Zylindermantels parallel laufen, gebildet, und die geladenen Partikel durchlaufen nacheinander die Potentialfelder bei I, II, III, IV usw. Im folgenden werden die durch die Knickstellen der Lecherleitung bewirkten Verzerrungen des Feldes als unwesentlich betrachtet, und es wird angenommen, daß die geladenen Teilchen sich im Innern jedes Zylinders mit konstanter Geschwindigkeit bewegen. Die Frequenz für die stehenden Wellen wird so gewählt, daß die Elektronen (Ionen) einen Zylinder gerade während der Dauer einer Periode (oder eines Vielfachen hiervon) durchlaufen. Die Spannung zwischen den Leitern wird deshalb, wie in der Figur gezeigt, jedesmal, wenn die Elektronen ein Potentialfeld passieren, die gleiche sein, und die Teilchen werden deshalb jedesmal mit einer vollen Wechselspannung beschleunigt. Da die Geschwindigkeit der Teilchen mit steigender Spannung zunimmt, werden die Zylinder mit fortschreitendem Abstand von der Teilchenquelle immer länger gemacht, und zwar proportional mit

$$v = c \ \frac{\sqrt{U^2 + 2U\varepsilon}}{U + \varepsilon} \qquad (1)$$

wobei

$$\varepsilon = \frac{m_0 c^2}{e} \qquad (2)$$

c = Lichtgeschwindigkeit, v = Geschwindigkeit der Teilchen, m_0 = Ruhemasse der Teilchen, e = Ladung der Teilchen, U = kinetische Spannung der Teilchen.

Die Wellenlängen der stehenden Wellen der Energieleitung sollen folglich auch nacheinander größer gemacht werden. Damit die Elektronen ständig mit der maximalen Wechselspannung beschleunigt werden, müssen die Wellenlänge in Vakuum λ_v und die Länge der stehenden Welle λ_{st} folgende Resonanzbedingungen erfüllen:

$$\frac{p}{4} \lambda_{st} = \frac{v}{c} \cdot \lambda_v \cdot \frac{q}{2} = l \qquad (3)$$

wobei l der Abstand zwischen zwei aufeinanderfolgenden Potentialfeldern ist, p und q zwei ganze Zahlen sind, von denen p angibt, wieviel Viertel der stehenden Welle man zwischen den Potentialfeldern hat, und q wie oft die Wechselspannung ihr Vorzeichen wechselt, während die Elektronen von einem Potentialfeld zum anderen gelangen. Dabei darf q nur dann eine ungerade Zahl sein, wenn p durch 4 teilbar ist, da andernfalls abwechselnde Beschleunigungen und Verzögerungen erhalten würden. Man

wird p und q so wählen, daß man günstige Werte für λ_v und l erhält, d. h. λ_v nicht zu klein, damit die Frequenz nicht zu groß wird, und l nicht zu groß, damit die Anordnung nicht zu lang wird. Dabei ist zu beachten, daß λ_{st} höchstens so groß wie λ_v werden kann. Will man z. B. Elektronen, deren Geschwindigkeit ungefähr gleich der Lichtgeschwindigkeit ist, beschleunigen, so kann man $p = 6$ und $q = 2$ wählen, und man erhält dann

$$\lambda_{st} = \frac{2}{3} \lambda_v \frac{v}{c} = \frac{2}{3} l \qquad (4)$$

Wenn die Geschwindigkeit der Teilchen stets kleiner ist als die halbe Lichtgeschwindigkeit, kann man $p = q = 2$ wählen und erhält

$$\lambda_{st} = 2 \lambda_v \frac{v}{c} = 2 l \qquad (5)$$

Dies entspricht der in Fig. 2 gezeigten Anordnung. Die Wellenlänge für die stehenden Wellen kann in bekannter Weise durch Anbringen von Materialien zwischen den Leitern, deren Dielektrizitätskonstante und/oder Permeabilität größer als 1 ist, der wachsenden Geschwindigkeit angepaßt werden. Um besonders kurze stehende Wellen zu erreichen, ohne die Frequenz zu sehr zu erhöhen, kann man Materialien mit hoher Dielektrizitätskonstante, wie z. B. Polystyrol, Polyäthylen, vorzugsweise in der Nähe der Spannungsmaxima (Spannungsbäuche) anbringen, während Materialien mit hoher Permeabilität vorzugsweise in der Nähe der Knotenpunkte der Spannung (Spannungsknotenpunkte) angebracht werden sollen. Diese Materialien sollen zweckmäßig möglichst kleine Wechselstromverluste haben. Dasselbe gilt auch für die Energieleitung überhaupt, die unter anderem mit kleinen Strahlungsverlusten gebaut werden soll. Man kann die Länge der stehenden Welle auch mit Hilfe von parallel geschalteten Kapazitäten bzw. seriegeschalteten Induktivitäten verändern. Um Absorptionsverluste zu verhindern, wird man die Elektronen (Ionen) in hohem Vakuum beschleunigen.

Die in Fig. 2 gezeigte Anordnung ist zur Beschleunigung von relativ langsamen Partikeln (Ionen unter 10 MV) besonders geeignet, da man bei großen Partikelgeschwindigkeiten lange Zylinder oder sehr hohe Frequenzen erhält.

Fig. 3 zeigt eine Anordnung, die sich für die Beschleunigung sehr schneller Partikel mit ungefähr Lichtgeschwindigkeit eignet, $\left(\frac{v}{c} \approx 1 \right)$. In diesem Fall sind die Beschleunigungszylinder so kurz gewählt, daß die Wechselspannung in der Zeit, die die Elektronen brauchen, um sich durch die Zylinder hindurch zu bewegen, sich nicht viel verändert, wobei die Elektronen allerdings nicht mit der maximalen Wechselspannung beschleunigt werden. Macht man z. B. die Zylinder 20 cm lang und wählt $\lambda_v = 6$ m, d. h. 30mal so lang ($q = 1/15$), so kann man bei $\lambda_{st} = 40$ cm ($p = 2$) fünf Beschleunigungsröhren anwenden, ohne daß die Phase der stehenden Welle insgesamt sich um mehr als 360°/6 = 60° ändert, d. h. daß die Beschleunigungsspannungen nicht weniger als cos ($\pm 30°$) = 86,7% der Maximalspannung betragen.

Wenn die Elektronen die fünf Beschleunigungsröhren durchlaufen haben, können sie noch weiter mit einer Wechselspannung, die um 60° in bezug auf die erste phasenverschoben ist, beschleunigt werden, und auf diese Weise kann man mit Hilfe eines Dreiphasenhochfrequenzsystems u_R, u_S, u_T die Elektronen in ununterbrochener Reihe beschleunigen. Nach einem aus fünf Röhren bestehenden Abschnitt, in dem sechs Beschleunigungen erfolgen, wird die Spannung u_R durch die Leitung L weitergeführt, auf der sich ebenfalls eine stehende Welle ausbildet. Die Phase am Ende dieser Leitung, die zu einem weiteren Beschleunigungsabschnitt führt, ist gegenüber dem Anfang um 180° gedreht. Wenn man den Mittelwert der Beschleunigungsspannungen mit u bezeichnet, so ist die kinetische Spannung am Ende des in Fig. 3 dargestellten Teiles der Anordnung $U_0 + 19\,u$. Um eine höchstmögliche Spannung bei einer bestimmten gesamten Röhrenlänge zu erreichen, sollen die Beschleunigungsröhren kurz gemacht werden. Wenn die Frequenz der Wechselspannung nicht zu hoch werden soll, soll darum das Verhältnis $\frac{\lambda_v}{\lambda_{el}}$ hoch gemacht werden, d. h. man soll Materialien mit sehr hoher Dielektrizitätskonstante und Permeabilität verwenden. Das Verhältnis kann auf bis über 15 ($\varepsilon\mu < 225$) gebracht werden und möglicherweise bis auf etwa 35 bis 40, unter Verwendung rutilhaltiger Dielektrika, z. B. Bariumtitanat usw. Die Beschleunigungsröhren können darum bei einer Frequenz von 50 MHz 20 cm lang gemacht werden $\left(\frac{\lambda_v}{\lambda_{el}} = 15\right)$, und es sollte möglich sein, eine Beschleunigung von 1 bis 2 MV pro Meter Apparatelänge zu erreichen.

Wenn die Geschwindigkeit der Teilchen wesentlich kleiner ist als die Lichtgeschwindigkeit, kann die in Fig. 3 gezeigte Anordnung ebenfalls angewendet werden, aber man wird, mit unveränderten Werten für die beiden Wellenlängen, eine entsprechend kleinere Anzahl Zylinder pro Phase verwenden müssen. Wenn $v = 0,2\,c$ ist, erhält man somit nur eine Beschleunigungsröhre pro Phase, wobei $q = {}^1/_3$ und die Beschleunigungsspannung $86,7\,^0/_0$ der Maximalspannung ist. Dabei erfolgt am Ende einer Röhre wie am Beginn der nächsten eine Beschleunigung mit dieser Spannung. Wenn die Geschwindigkeit der Teilchen noch kleiner ist, hat das zur Folge, daß die Beschleunigungsspannung weiter sinkt. Wenn $v = 0,133\,c$ und $\frac{\lambda_v}{\lambda_{el}} = 15$ (somit $q = {}^1/_2$) ist, erhält man mit einer Beschleunigungsröhre pro Phase und einem Zweiphasensystem eine Beschleunigungsspannung, die $\cos(\pm 45°) = 70,7\,^0/_0$ der maximalen Wechselspannung beträgt. Hieraus ersieht man, daß das Anwendungsgebiet für die in Fig. 3 gezeigte Anordnung sich an das Gebiet anschließt, wo es vorteilhaft sein wird, die in Fig. 2 gezeigte Anordnung zu verwenden.

Wenn man eine große Apparatelänge zu vermeiden wünscht (100 MV-Deuteronen würden bei einer Beschleunigungsspannung von 200 kV bei 50 MHz [$\lambda_v = 6$ m] und $\frac{\lambda_v}{\lambda_{el}} = 15$ eine Apparatelänge von 110 bis 120 m erfordern), kann, man mit magnetischen Steuerfeldern den Teilchen eine Kreisbewegung erteilen und sie dazu bringen, viele Male eine oder mehrere Beschleunigungsröhren zu durchlaufen.

Da die Umlaufzahl der Teilchen sehr groß gemacht werden kann, hat das auch den Vorteil zur Folge, daß die Beschleunigungsspannung bedeutend kleiner als bei der geradlinigen Anordnung gehalten werden kann, z. B. etwa 10 kV.

Fig. 4 zeigt eine Anordnung, die insbesondere für die Beschleunigung von Elektronen geeignet ist. Es sind in diesem Fall zwei Energieleitungen vorhanden, die durch die Leiter 1, 2 bzw. 5, 6, 7 und 4, 3 bzw. 6, 8, 9 gebildet werden. λ_{el} ist gleich $^2/_3$ des Umfanges $2\pi R$ gewählt worden ($p = 6$). Man erhält dann zwei Spannungsknotenpunkte bei 10 und 10^a, wo die Leiter kurzgeschlossen sind, während die Punkte 11 und 12 sich in der Nähe eines Spannungsbauches befinden. Auf der Strecke 10, 16, 10^a wird die Hochfrequenzspannung Null sein, und man kann deswegen die kreisförmigen Leiter alle beispielsweise bei 16 unterbrechen. Man vermeidet damit, daß das variierende Magnetfeld 15 in den Kreisleitern Ströme induziert. Man kann übrigens auch die beiden Energieleitungen in den Knotenpunkten 10 und 10^a abschalten und das ganze dazwischenliegende Stück der Beschleunigungsröhre, das von der Energieleitung isoliert sein kann, an Erdpotential legen. Aus diesem Grunde braucht die Länge des hohlen, auf Erdpotential sich befindenden Leiterstückes 2 bzw. 3 auch nicht unbedingt ein ganzzahliges Vielfaches p von $\frac{\lambda_{el}}{4}$ zu sein, was dagegen nötig ist, wenn man die Elektronen in mehreren im Kreise angeordneten Potentialfeldern beschleunigen will, statt in einem einzigen Feld zwischen den Punkten 11 und 12.

Es ist auch nicht notwendig, wie in Fig. 4 gezeigt, zwei Energieleitungen zu verwenden, die um 180° phasenverschoben sind; man kann auch, wie in Fig. 5 gezeigt, eine einfache Energieleitung verwenden, bei welcher $^1/_4$ der Wellenlänge der stehenden Welle dem Teil 19, 20 der Beschleunigungsröhre 22, 23 entspricht, aber der übrige Teil der Beschleunigungsröhre vom Knotenpunkt 18 der Welle bis zum Potentialfeld 17 mit Erde verbunden und elektrisch von der Wechselspannung getrennt ist. Mit 25 ist die Elektronenspritze bezeichnet, die sich im Rohr 26 befindet und die Elektronen bei 27 in die Kreisröhre 22, 23 einspritzt, d. h. an einer Stelle, wo die Hochfrequenzspannung Null ist.

Fig. 4 zeigt, daß die Energieleitung über einen Transformator 13 an den Hochfrequenzgenerator 14 geschaltet ist; dies ist aber nicht wesentlich. Die Leiter 2 und 3 bilden eine Kreisröhre (Toroidröhre), die zwischen den Elektroden 11 und 12 offen ist, und in dieser Röhre, die um die Ausbildung von Wirbelströme durch das magnetische Steuerfeld zu verhindern, der Länge nach aufgeschlitzt ist. Von parallelen isolierten Leitern gebildet ist, zirkulieren die Elektronen. Um die Zentrifugalkraft aufzuheben, ist ein magnetisches Feld senkrecht zur Papierebene angebracht. Dieses erzeugt Lorenzkräfte auf die zirkulierenden Elektronen. Man erreicht eine sowohl radial wie auch senkrecht dazu gerichtete Stabilisierung der Elektronenbahnen dadurch, daß man das

4   847 318

Magnetfeld in der Richtung des Radius R abnehmen läßt, aber schwächer wird, proportional zu R_r^{-1}. Die Elektronen werden durch die Stabilisierungskräfte nach der kreisförmigen Röhrenachse hingedrängt.

5 Wenn die Geschwindigkeit des Elektrons zunimmt, nimmt seine Masse auch zu, und das Magnetfeld muß zunehmen, um die erhöhte Zentrifugalkraft aufzuheben. Man kann zu diesem Zweck ein magnetisches Wechselfeld mit verhältnismäßig niedriger

10 Frequenz $\left(z. B. \dfrac{\omega}{2\pi} = 50 \text{ Hz}\right)$ verwenden, um die Elektronen zu steuern. Damit die Elektronen immer derselben Kreisbahn mit dem Radius R folgen, muß gemäß der bekannten Theorie der Strahlentransfor-

15 matoren der Zusammenhang zwischen der Feldstärke B des Steuerfeldes und der Elektronenspannung U der folgende sein:

$$B = \frac{1}{cR}\ \sqrt{U^2 + 2U\varepsilon} \sim \frac{U}{cR} \text{ (wenn } U \gg \varepsilon) \quad (6)$$

20 Das Steuerfeld muß deshalb ungefähr proportional mit der Elektronenspannung zunehmen. Wenn die Umlaufzeit Δt der Elektronen und der Radius R als konstant angenommen werden (was natürlich nur für einen kurzen

25 Abschnitt der Beschleunigungsperiode zulässig ist, und nur weil die Geschwindigkeit langsamer wächst als U), ist die Zunahme ΔU der Elektronenspannung in der Zeit Δt gleich der beschleunigenden Hochfrequenz-

30 spannung u. Daher ist $u = \Delta U = \dfrac{dU}{dt}\ \Delta t$, d. h. proportional $\dfrac{dU}{dt}$ und damit proportional $\dfrac{dB}{dt}$. Die Amplitude der Hochfrequenzspannung muß also proportional mit $\cos \omega\ t$ abnehmen, d. h. entsprechend

35 mit der Niederfrequenz ω moduliert sein, wenn das Steuerfeld proportional mit $\sin \omega\ t$ zunimmt. Bei konstanter Hochfrequenzspannung muß das Steuerfeld dagegen proportional mit der Zeit t zunehmen. Wenn das Magnetfeld während des kurzen Zeitabschnittes lang-

40 samer zunimmt, als dem Proportionalitätsfaktor entspricht, so werden die Elektronen bei konstanter Hochfrequenzspannung eine zu hohe kinetische Spannung erhalten, und der Radius der Elektronenbahn wird zunehmen. Da die Geschwindigkeit weniger zunimmt als der Um-

45 fang des Kreises, werden die Elektronen das Potentialfeld etwas nach dem Maximum der Hochfrequenzspannung erreichen. Die Phasenverspätung wird sich bei jedem Umlauf vergrößern und bewirken, daß die Beschleunigungsspannung abnimmt. Das wird so lange vor sich

50 gehen, bis die Elektronenbahn sich in dem betrachteten kurzen Abschnitt der Beschleunigungsperiode genau auf die richtige Beschleunigungsspannung eingespielt hat, die dem Werte von $\dfrac{dB}{dt}$ während dieses Zeit-

55 abschnittes entspricht. Man erhält somit jeweils einen stabilen Gleichgewichtsradius für die Elektronenbahn. Analoges gilt, wenn die Amplitude einer Hochfrequenzspannung, die mit $\cos(\omega\ t)$ moduliert ist, beim Beginn der Beschleunigungsperiode genügend groß ist.

60 Wenn die Elektronen in den Beschleuniger mit einer Anfangsspannung von beispielsweise 460 kV eintreten, werden sie etwa 85% der Lichtgeschwindigkeit besitzen. Die Geschwindigkeit wird nachträglich

praktisch bis zur Lichtgeschwindigkeit (bei 10 MV ist die Differenz nur etwa $0,23\%$), zunehmen. Damit 65 die Umlaufzeit auf alle Fälle eine Hochfrequenzperiode sei, muß man entweder die Radien der Elektronenbahnen oder auch die Frequenz während der Beschleunigungsperiode ändern. Wählt man das erstere, so muß dem Steuerfeld und der Beschleuni- 70 gungsröhre eine so große Ausdehnung in radialer Richtung gegeben werden, daß man die nötige Vergrößerung der Elektronenbahn, z. B. von 0,85 bis zu 1, zulassen kann. Da ein konstruktiv günstig wäre, ein möglichst schmales Magnetfeld zu erhalten, sollte die 75 Anfangsspannung der Elektronen so hoch wie möglich sein. Es wird daher konstruktiv günstig sein, die Elektronenspritze außerhalb der Beschleunigungsröhre anzuordnen und die Elektronen auch in der Elektronenspritze mit hochfrequenten Feldern von derselben 80 Frequenz wie in der Beschleunigungsröhre zu beschleunigen. Die Elektronen können dann mit etwas zu großer Spannung eingeführt werden, so daß sie die Innenwand der Beschleunigungsröhre streifen müssen. Wenn man hier einige kurze und dünne 85 Bremsfolien anbringt, können die Elektronen so viel abgebremst werden, daß sie gerade die Spannung erhalten, welche der innersten Elektronenbahn entspricht. Mit wachsender Spannung wächst der Radius der Bahn, und die Elektronen werden nicht mehr durch 90 die Bremsfolien gestört. Da man die Elektronen nur während eines kleinen Teiles der Hochfrequenzperiode einführen kann, sollten die Elektronen nur während eines gewissen Teiles der Periode emittiert werden. Die Elektronen können vorzugsweise in den Knoten- 95 punkten der Beschleunigungsspannung eingeführt werden bzw. dort, wo die Energieleitung nicht vorhanden ist.

Die umlaufende azimutal abgegrenzte Elektronenladung wird eine schwache Wechselspannung in- 100 duzieren, wenn sie einen Kondensator, der im Beschleunigungsrohr am Knotenpunkt der Spannung angebracht ist, passiert. Diese Wechselspannung kann verstärkt und für die Steuerung der Hochfrequenzgeneratoren gebraucht werden. Dies ist besonders 105 wichtig, wenn man die Frequenz des Generators während der Beschleunigungsperiode ändern will, um die Resonanzbedingung $\dfrac{v}{c} \cdot \dfrac{q}{\ }\ \lambda_r = 2\ R\pi$ (q = gerade ganze Zahl) zu erfüllen. In diesem Fall sollen auch 110 die Konstanten der Energieleitung verändert werden, damit λ_r dem Abstande 11 bis 10 gleich bleibt. Dies kann z. B. durch die Benutzung von Induktivitäten mit Eisenkernen oder ferromagnetischen Materialien zwischen den Leitern, deren Permeabilität mit Hilfe 115 einer variablen Gleichstromvormagnetisierung verändert wird, geschehen.

Es wird wesentliche konstruktive Vorteile bieten, den Steuerfluß durch die Flächen innerhalb der Elektronenbahn zu schließen und die Magnetisierungs- 120 wicklung um den Eisenkern, der dabei gebildet wird, anzubringen. Der Kernfluß ist also entgegengesetzt gerichtet wie der Steuerfluß, d. h. umgekehrt wie bei einem Strahlentransformator. 125

Das wird jedoch zur Folge haben, daß der variierende Kernfluß ein elektrisches Wirbelfeld erzeugen wird, das die Bewegung der Elektronen abzubremsen sucht. Da der Kernfluß bei Annahme konstanter Induktion im Eisenkern nur einen Teil, z. B. die Hälfte der Kreisfläche ausfüllen wird, (wenn die Breite des Steuerfeldes αR_0 ist, wird die Fläche des Kernflusses etwa 2α mal so groß wie die Kreisfläche πR_u^2), und da die Induktion im Kern nur $1/2$ so groß ist wie im Kern eines Strahlentransformators, wird die induzierte Gegenspannung nur z. B. $1/4$ der kinetischen Spannung sein, die dem Steuerwechselfeld entspricht. Die Beschleunigerspannung muß in diesem Fall entsprechend, d. h. 25% größer als ohne bremsendes Wirbelfeld gemacht werden.

Wenn die Elektronen die gewünschte Geschwindigkeit erreicht haben, können sie aus dem Beschleunigungsprozeß durch plötzliches Einschalten eines dem Steuerfeld überlagerten magnetischen Zusatzfeldes (positiv oder negativ) herausgebracht werden. Auf gleiche Weise wie in einem Strahlentransformator können die Elektronen in einer Antikathode zur Erzeugung von γ-Strahlen abgebremst werden oder auch mit Hilfe von besonderen Ablenkungselektroden aus der Beschleunigungsröhre herausgeführt werden.

Um eine günstige Einführung der Elektronen zu erreichen, kann es auch vorteilhaft sein, ein magnetisches Zusatzfeld zu verwenden, welches plötzlich eingeschaltet wird und die Elektronen von den früher genannten Bremsfolien oder von anderen reellen oder fiktiven Kathoden, welche den Umlauf behindern können, entfernt.

Mit Bezug auf Fig. 5 soll noch erwähnt werden, daß man beispielsweise bei einem Elektronenbahnradius von $R_0 =$ etwa $1,5$ m und einer maximalen Steuerfeldinduktion von etwa $11\,000$ Gauß eine maximale kinetische Spannung von etwa 500 MV für die Elektronen erreichen könnte. Die Beschleunigerfrequenz sollte etwa 32 MHz (Wellenlänge $\lambda_r = 2\pi R_0 = 9,4$ m) sein, wobei eine Frequenz von 50 Hz für

$$U = \sqrt{\varepsilon^2 + c^2 B_t' R^2} - \varepsilon \sim \frac{B_t' c^2 R^2}{2\varepsilon} = \frac{c^2 B_{t}'^2 R_u^{2K}}{2\varepsilon} R^{2(1-K)} \tag{8}$$

$$R = \left[\frac{2 U \varepsilon}{c^2 B_{t}'^2 R_u^{2K}}\right]^{\frac{1}{2(1-K)}} \tag{9}$$

$$v = \frac{c^2 B_{0t}}{\varepsilon} R_u^K R^{(1-K)} \tag{10}$$

$$\nu = \frac{c^2 B_{0t}}{2\pi\varepsilon} R_u^K R^{-K} \tag{11}$$

Für Protonen ist $\varepsilon = 930$ MV und für Deuteronen 1860 MV.

Wenn die Umlauffrequenz konstant gleich

$$\nu_0 = \frac{c^2 B_0}{2\pi\varepsilon} \tag{12}$$

Wenn der Bahnradius als Folge der Spannungserhöhung sich dem größten Wert, der konstruktiv

das Steuerfeld (Beschleunigungszeit $=$ max. $1/200$ Sek.) eine maximale Beschleunigungsspannung von etwa 10 kV erforderlich machen würde. Diese Zahl zeigt, daß man mit Hilfe der beschriebenen Anordnung mit technisch angemessenen Mitteln eine höhere Spannung erzeugen kann, als mit irgendeinem anderen bis jetzt bekannten Apparat, Strahlentransformatoren inbegriffen.

Wenn man Ionen nach dem in Fig. 4 benutzten Prinzip beschleunigen will, wird das Geschwindigkeitsintervall für die Teilchen so groß werden, daß man die Resonanzbedingung mit einer konstanten Beschleunigungsfrequenz durch Verändern der Bahnradien nicht erfüllen kann. Es wird auch große Schwierigkeiten bieten, die Beschleunigungsfrequenz und die Konstanten der Energieleitung innerhalb des nötigen Bereiches zu verändern. Man kann in diesem Fall, wo $\frac{v}{c}$ (in Gleichung 3) klein ist, einen entsprechend großen Wert für q wählen und somit die Umlauffrequenz der Ionen mit einem Bruchteil der Beschleunigungsfrequenz synchronisieren. Wenn die Geschwindigkeit der Ionen zunimmt, wird, wie früher erwähnt, der Bahnradius zunehmen und die Ionen werden automatisch die Resonanzbedingung erfüllen und an dieser untersynchronen Bewegung festhalten. Wenn die Feldstärke B_t des magnetischen Steuerfeldes zur Zeit t als Funktion des Radius R nach Gleichung

$$B = B_0\left(\frac{R}{R_0}\right)^{-K} \tag{7}$$

wo $0 < K < 1$, abnimmt, werden die Spannung U der Ionen, der Bahnradius r, die Geschwindigkeit v und die Umlauffrequenz ν durch folgende Gleichungen (8) bis (11) bestimmt, die für das nicht relativistische Gebiet ($U \ll \varepsilon$) gelten, und sich leicht aus der Gleichung (6) ableiten lassen:

sein soll, gibt dies die folgenden Gleichungen zwischen Magnetfeld B_{0t}, das sich nur mit der Zeit verändert, und dem Bahnradius R bzw. der Spannung U

$$R = R_0\left(\frac{B_{0t}}{B_0}\right)^{\frac{1}{K}} \tag{13}$$

$$U = \frac{c^2 B_u^2 R_u^2}{2\varepsilon}\left(\frac{B_{0t}}{B_0}\right)^{2K} \tag{14}$$

Wenn das magnetische Steuerfeld sich während eines Teilchenumlaufes $\Delta t = \frac{2\pi R}{v} = \frac{1}{\nu_0}$ von B_0 auf $B_{0t} = B_0 + \frac{dB_0}{dt}\Delta t$ erhöht, so ergibt sich mit Gleichung (14) für die nötwendige Beschleunigungsspannung pro Umlauf:

$$u = U_0 - U_{\Delta t} = \frac{c^2 B_u^2 R_u^2}{2}\left[\left(1 + \frac{1}{B_0}\frac{dB_0}{dt}\Delta t\right)^{2K} - 1\right] = \frac{2 R_u^2}{K}\frac{dB_0}{dt} \tag{15}$$

möglich ist, nähert, sollen die Ionen mit Mitteln, die später beschrieben werden, aus dem Synchronismus herausgebracht werden. Bei der nun folgenden asynchronen Bewegung werden die Ionen im Mittel nicht beschleunigt, und als Folge der Erhöhung des Steuerfeldes wird der Bahnradius darum abnehmen.

5 Dies setzt sich so lange fort, bis die Ionen, deren Umlauffrequenz wegen der Abnahme der Bahnradien ständig zunimmt, eine höhere untersynchrone Frequenz (wobei $q_2 < q_1$) erreichen und sich synchronisieren, so daß sie wieder beschleunigt werden können.

10 Auf diese Weise wird sich das Spiel fortsetzen, bis die Ionen ihre maximale Geschwindigkeit erreicht haben. Die Beschleunigungsfrequenz soll so hoch gewählt werden, daß ein relativer Unterschied zwischen den beiden letzten untersynchronen Frequenzen

15 $\left(\text{somit } \dfrac{q_{n-1} - q_n}{q_n} \right)$ kleiner wird als der Unterschied zwischen dem größten und kleinsten Bahnradius. Wenn man Deuteronen bis 100 MV beschleunigen will, so ist die maximale Geschwindigkeit, die man

20 aus der Gleichung (1) errechnen kann (ε für Deuteronen = 1860 MV) etwa 0,315 c. Wenn das magnetische Steuerfeld für den größten Bahnradius maximal etwa 11 000 Gauß ist, wird der größte Bahnradius etwa 1,9 m werden. Wenn man bei einer

25 Beschleunigungsfrequenz von 39,5 MHz, d. h. = 7,6 m, $\lambda_{el} = \lambda$, (d. h. $\varepsilon\mu = 1$) macht, wird bei $p = 6$ entsprechend Fig. 4 der Abstand zwischen zwei Spannungsmaxima 11,4 m, die maximale Spannung wird somit ziemlich genau bei den Beschleunigungs-

30 elektroden der Beschleunigungsröhre liegen. Der kleinste mögliche Wert von q ist 10, der nächst größere, bei dem die Resonanzbedingung wieder erfüllt ist, ist 12. Der kleinste Bahnradius wird somit (12-10)/10 = 20% kleiner als der größte, d. h.

35 etwa 1,58 m werden. Wenn der Frequenz des Steuerfeldes 50 Hz und $K = \frac{2}{3}$ beträgt, wird die maximale Synchronspannung, wie man aus der Gleichung (15) errechnen kann, $u = 11,8$ kV $\Big($ bei einer Maximal-

40 induktion von 11 000 Gauß und $f = 50$ Hz ergibt sich im Anfang $\dfrac{dB_0}{dt}$ zu $\omega B_m = 314 \cdot 11000 \cdot 10^{-8} \dfrac{\text{V}}{\text{cm}^2}$

45 $= 0{,}0345 \dfrac{\text{V}}{\text{cm}^2} \Big)$.

Um die Ionen aus dem Synchronismus herauszubringen, wenn der Frequenzwechsel stattfinden soll, kann auf mehrere Arten vorgegangen werden. Man kann zu bestimmten Zeiten auf bekannte Weise die

50 Beschleunigungsfrequenz etwas verändern. Man kann auch auf bekannte Weise die Beschleunigungsspannung ändern und deren Wert unter ein der früher berechneten (Gleichung 15) Synchronwert sinken lassen. Die Modulationsfrequenz müßte in diesem Fall

55 ungefähr nach einer e^{-t}-Funktion abnehmen, und die Zeitintervalle, in denen die Teilchen sich synchron bzw. asynchron bewegen (die angenähert gleich groß sein werden), müßten somit so abgepaßt sein, daß die Bahnradien nicht die zulässigen äußeren und

60 inneren Grenzen überschreiten. Man kann den Synchronismus auch dadurch aufheben, daß man die notwendige Synchronspannung u über die vorhandene Beschleunigungsspannung erhöht, die man als pro-

portional mit $\dfrac{dB}{dt}$, d. h. proportional cos ωt sich

65 ändernd annimmt. Dies kann z. B. dadurch geschehen, daß man das Steuerfeld ändert und somit $\dfrac{dB}{dt}$ periodisch erhöht, doch ohne die Beschleunigungsspannung entsprechend zu ändern. Eine einfachere Lösung wird

70 sein, die Form der Pole des Steuerfeldes derart zu verändern, daß das Steuerfeld weniger stark abnimmt, d. h. K wird kleiner, wenn man den größten Bahnradius erreicht. Durch Verkleinerung von K von

z. B. $\frac{2}{3}$ auf $\frac{1}{3}$ (in Gleichung 15) wird die Synchron-

75 spannung auf den doppelten Wert steigen, was genügen wird, um den Synchronismus aufzuheben. Dasselbe kann auch dadurch erreicht werden, daß man den Beschleunigungselektroden eine solche Form gibt, daß die Richtung des Potentialfeldes sich

80 ändert und die longitudinale Feldkomponente kleiner wird beim größten Bahnradius. Auf diese Weise kann die Beschleunigungsspannung kleiner gemacht und unter den Synchronwert gebracht werden, wodurch der Synchronismus aufgehoben wird. Damit diese

85 geometrisch bedingten Lösungen, die auch gleichzeitig benutzt werden können, angewendet werden können, darf die Beschleunigungsspannung normalerweise die Synchronspannung nicht um mehr als z. B.

90 30% überschreiten und muß daher, wie früher erwähnt, proportional $\dfrac{dB}{dt}$ geändert werden.

Die hier angegebenen Methoden können selbstverständlich auch für die Beschleunigung von Elek-

95 tronen benutzt werden, wenn die Anfangsgeschwindigkeit so klein ist, daß die Geschwindigkeitszunahme die Ausdehnung des Steuerfeldes in radialer Richtung überschreitet.

Eine Anordnung der beschriebenen Art für 100 MV-

100 Deuteronen mit einem größten Bahnradius von etwa 1,9 m wird weniger als 130 t wiegen und würde sich somit bedeutend günstiger stellen als ein entsprechendes Zyklotron mit einem Gewicht von über 5000 t. Es zeigt sich somit, daß man mit der beschrie-

105 benen Einrichtung mit geringerem Aufwand auch Ionen auf wesentlich höhere Spannungen als mit bis jetzt bekannten Apparaten beschleunigen kann.

PATENTANSPRÜCHE: 110

1. Anordnung zur Beschleunigung von elektrisch geladenen Teilchen mit Hilfe von in der Bewegungsrichtung der Teilchen aufeinanderfolgenden hochfrequenten elektrischen Potentialfeldern, dadurch gekennzeichnet, daß die Elektroden, 115 zwischen denen die Potentialfelder erzeugt werden, hohle Stücke der beiden Leiter einer Hochfrequenzenergieleitung sind, auf der stehende Spannungswellen erzeugt werden, wobei die geladenen Teilchen, nachdem sie ein Potentialfeld durch- 120 laufen haben, sich innerhalb eines der genannten hohlen Stücke bewegen.

2. Anordnung nach Anspruch 1, dadurch gekennzeichnet, daß je zwei aufeinanderfolgende Potentialfelder einen Abstand voneinander be- 125 sitzen, der den vierten Teil oder ein Vielfaches

davon der Wellenlänge der zwischen diesen Feldern stehenden Spannungswelle beträgt.

3. Anordnung nach Anspruch 2, dadurch gekennzeichnet, daß die Geschwindigkeit der geladenen Teilchen und die Abstände zwischen den Potentialfeldern so einander angepaßt sind, daß diese Abstände mit einem Zeitunterschied durchlaufen werden, der gleich der Hälfte der Hochfrequenzperiode ‚oder einem Vielfachen davon ist.

4. Anordnung nach Anspruch 2, dadurch gekennzeichnet, daß die Energieleitung mit parallel geschalteten Kapazitäten und seriengeschalteten Induktivitäten versehen ist, um die jeweilige Wellenlänge der stehenden Wellen und damit die Abstände der Potentialfelder der Teilchengeschwindigkeit anzupassen.

5. Anordnung nach Anspruch 1, dadurch gekennzeichnet, daß der Raum zwischen den Leitern mindestens teilweise durch Materialien, deren Dielektrizitätskonstante und Permeabilität größer als 1 ist, ausgefüllt ist, wobei diese Materialien kleine Hochfrequenzverluste haben, um ein jeweilige Wellenlänge der stehenden Wellen und damit die Abstände der Potentialfelder der Teilchengeschwindigkeit anzupassen.

6. Anordnung nach Anspruch 5, dadurch gekennzeichnet, daß die Materialien hoher Permeabilität hauptsächlich an den Spannungsknotenpunkten, die Materialien hoher Dielektrizitätskonstante dagegen hauptsächlich an den Spannungsmaxima angebracht sind.

7. Anordnung nach Anspruch 6, dadurch gekennzeichnet, daß der Abstand zwischen den Potentialfeldern die Hälfte der Länge der stehenden Welle beträgt und daß die Wellenlänge der Hochfrequenzspannung in Vakuum im Verhältnis hierzu so groß ist, daß die Hochfrequenzspannung sich in der Zeit, welche die geladenen Teilchen brauchen, um einen oder mehrere Abstände der Potentialfelder zu durchlaufen, nur wenig ändert.

8. Anordnung nach Anspruch 7, dadurch gekennzeichnet, daß die geladenen Teilchen nach ihrer Beschleunigung durch eine Hochfrequenzspannung in zwei oder mehreren Potentialfeldern auf gleiche Weise in Feldern beschleunigt werden, die von im Verhältnis zu der ersten Wechselspannung phasenverschobenen Hochfrequenzspannungen erzeugt sind, indem diese Spannungen ein symmetrisches Mehrphasensystem bilden, wobei die geladenen Teilchen dann beschleunigt werden, wenn die Felder nicht stark von ihrem Maximalwert abweichen.

9. Anordnung nach Anspruch 1, dadurch gekennzeichnet, daß die genannten hohlen Leiterstücke einen Teil einer in sich geschlossenen Beschleunigungsröhre bilden, in der hohes Vakuum herrscht und durch welche die geladenen Teilchen mehrmals hindurchgeführt werden.

10. Anordnung nach Anspruch 9, dadurch gekennzeichnet, daß die Beschleunigungsröhre kreisförmig ist, wobei die geladenen Teilchen mit Hilfe

eines zeitlich veränderlichen magnetischen Steuerfeldes mehrmals durch dieselbe hindurchgeführt werden.

11. Anordnung nach Anspruch 10, dadurch gekennzeichnet, daß die Energieleitung kurzgeschlossen ist in einem längs der Beschleunigungsröhre gemessenen Abstand vom beschleunigenden Potentialfeld, der 1/4 der Wellenlänge der stehenden Wellen beträgt, so daß der übrige Teil der Beschleunigungsröhre von der Kurzschlußstelle bis zum Potentialfeld nicht von der Hochfrequenzspannung beeinflußt wird.

12. Anordnung nach Anspruch 10, dadurch gekennzeichnet, daß zwei Energieleitungen, deren Wechselspannungen um 180° zueinander phasenverschoben sind, vorgesehen sind, wobei jede der Energieleitungen von zwei Teilen der Beschleunigungsröhre gebildet wird, die vom Potentialfeld getrennt sind, und in einem längs der Beschleunigungsröhre gemessenen Abstand vom Potentialfeld kurzgeschlossen ist, der 1/4 der Länge der stehenden Wellen beträgt, während der restliche Teil der Beschleunigungsröhre zwischen den zwei Knotenpunkten nicht von den Hochfrequenzspannungen beeinflußt wird.

13. Anordnung nach Anspruch 10, dadurch gekennzeichnet, daß das genannte hohle Leiterstück der Länge nach an mehreren Stellen aufgeschlitzt ist.

14. Anordnung nach Anspruch 10, dadurch gekennzeichnet, daß das genannte hohle Leiterstück durch mehrere parallele, voneinander isolierte leitende Teile gebildet ist.

15. Anordnung nach Anspruch 10, dadurch gekennzeichnet, daß das magnetische Steuerfeld in radialer Richtung abnimmt, jedoch weniger stark als proportional zu R^{-1}, wobei R der Abstand von der Zentralachse ist.

16. Anordnung nach Anspruch 10, dadurch gekennzeichnet, daß das magnetische Steuerfeld derart verteilt und die Beschleunigungsröhre in radialer Richtung derart bemessen ist, daß der Radius der Teilchenbahnen im gleichen Verhältnis wie die Teilchengeschwindigkeit mit steigender Spannung wächst, und somit die Zeit für einen Umlauf des Teilchens immer konstant ist.

17. Anordnung nach Anspruch 16, dadurch gekennzeichnet, daß die Teilchen in den Beschleunigungsprozeß mit so hoher Anfangsspannung eingeleitet werden, daß die Geschwindigkeitszunahme während des Beschleunigungsprozesses nicht größer als 25% ist.

18. Anordnung nach Anspruch 10, dadurch gekennzeichnet, daß das magnetische Steuerfeld mit kleinerer Frequenz als 1000 Hz verändert wird und daß 1/4 jeder Periode zur Beschleunigung der Teilchen benutzt wird.

19. Anordnung nach Anspruch 10, dadurch gekennzeichnet, daß die Amplitude der Hochfrequenzspannung während der Beschleunigungszeit proportional mit $\frac{dB}{dt}$ variiert, wobei B die Feldstärke des magnetischen Steuerfeldes ist.

20. Anordnung nach Anspruch 10, dadurch gekennzeichnet, daß die Amplitude der Hochfrequenzspannung etwas größer ist als die Spannungszunahme, die der Zunahme des magnetischen Steuerfeldes während eines Umlaufes entspricht, und daß die Elektronen derart in den Beschleunigungsprozeß eingeführt werden, daß sie das Potentialfeld etwas später als beim zeitlichen Höchstwert passieren.

21. Anordnung nach Anspruch 10, dadurch gekennzeichnet, daß der Magnetfluß des Steuerfeldes sich durch die innere Öffnung der Beschleunigungsröhre schließt und auch die Magnetisierungswicklung in dieser Öffnung angebracht ist.

22. Anordnung nach Anspruch 10, dadurch gekennzeichnet, daß die Teilchen aus einer Quelle außerhalb der Kreisbahn mit etwas zu großer Spannung in die Kreisröhre eingeführt werden und dazu gebracht werden, die Innenwand dieser Röhre zu streifen, an der sie mit Hilfe von dünnen, radial gestellten, kurzen Bremsfolien, durch welche sie hindurchdringen, abgebremst werden, bis die richtige Anfangsspannung erreicht ist.

23. Anordnung nach Anspruch 10, dadurch gekennzeichnet, daß die Teilchen in demjenigen Teil der Beschleunigungsröhre eingebracht werden, an dem die Hochfrequenzspannung gleich Null ist.

24. Anordnung nach Anspruch 10, dadurch gekennzeichnet, daß die Anfangsspannung der Teilchen mit einer Spannung erzeugt wird, welche dieselbe Frequenz hat, wie die Potentialspannung erzeugende Hochfrequenzspannung, und daß die Teilchen periodisch während eines kleinen Teiles der Hochfrequenzperiode emittiert werden.

25. Anordnung nach Anspruch 24, dadurch gekennzeichnet, daß die periodisch durchlaufenden Teilchen einen Kondensator durchlaufen, der in der Beschleunigungsröhre an einer Stelle, wo die Hochfrequenzspannung Null ist, angebracht ist und in diesem Kondensator eine schwache Wechselspannung hervorrufen, die verstärkt wird und den Hochfrequenzgenerator für die Hochfrequenzspannung steuert.

26. Anordnung nach Anspruch 10, dadurch gekennzeichnet, daß die Bahnradien der Teilchen beinahe konstant sind, während die Beschleunigungsfrequenz proportional mit der Umlauffrequenz der Teilchen zunimmt.

27. Anordnung nach Anspruch 26, dadurch gekennzeichnet, daß die Energieleitung der Änderung in der Beschleunigungsfrequenz mit Hilfe von ferromagnetischem Material angepaßt ist, dessen Permeabilität durch variable Vormagnetisierung mit Gleichstrom verändert wird.

28. Anordnung nach Anspruch 10, dadurch gekennzeichnet, daß die Teilchen in bzw. aus dem Beschleunigungsprozeß gebracht werden mit Hilfe von variierenden Magnetfeldern, die dem Steuerfeld überlagert sind und die dann eingeschaltet werden, wenn die Beschleunigung der Teilchen beginnt bzw. aufhört.

29. Anordnung nach Anspruch 10, dadurch gekennzeichnet, daß die geladenen Teilchen sukzessive bei mehreren Umlauffrequenzen beschleunigt werden, die zunehmend Unterfrequenzen der Hochfrequenz sind, wobei die Synchronisierung der Frequenzen jeweils für kurze Zeit aufgehoben wird, wenn der Bahnradius einen gewissen Wert überschreitet.

30. Anordnung nach Anspruch 29, dadurch gekennzeichnet, daß der Synchronismus durch periodische Änderung der Beschleunigungsfrequenz aufgehoben wird.

31. Anordnung nach Anspruch 29, dadurch gekennzeichnet, daß der Synchronismus durch periodische Senkung des Maximalwertes der Beschleunigungsspannung unter den niedrigsten Wert, bei welchem Synchronismus möglich ist, aufgehoben wird.

32. Anordnung nach Anspruch 29, dadurch gekennzeichnet, daß das Steuerfeld periodisch verändert wird, und daß die zeitliche Ableitung der Feldstärke desselben kurzzeitig um so viel erhöht wird, daß die Synchronspannung die maximale Beschleunigungsspannung überschreitet.

33. Anordnung nach Anspruch 29, dadurch gekennzeichnet, daß die Steuerfeld von einem bestimmten Bahnradius an bei vergrößertem Radius langsamer abnimmt, wobei die Änderung so groß ist, daß die notwendige Synchronspannung größer wird als die maximale Beschleunigungsspannung.

34. Anordnung nach Anspruch 29, dadurch gekennzeichnet, daß die Elektroden für das beschleunigende Potentialfeld eine solche Form haben, daß die zum Bahnkreis tangentiale Feldkomponente abnimmt, wenn der Bahnradius einen gewissen Wert überschreitet, so daß die geladenen Teilchen weniger stark, als zur Aufrechterhaltung des Synchronismus notwendig ist, beschleunigt werden.

35. Anordnung nach Anspruch 29, dadurch gekennzeichnet, daß die Beschleunigungsspannung abgesehen von der genannten kurzen Zeit proportional zu der zeitlichen Ableitung der Feldstärke des Steuerfeldes ist.

Hierzu 1 Blatt Zeichnungen

⊕ 5308 8. 52

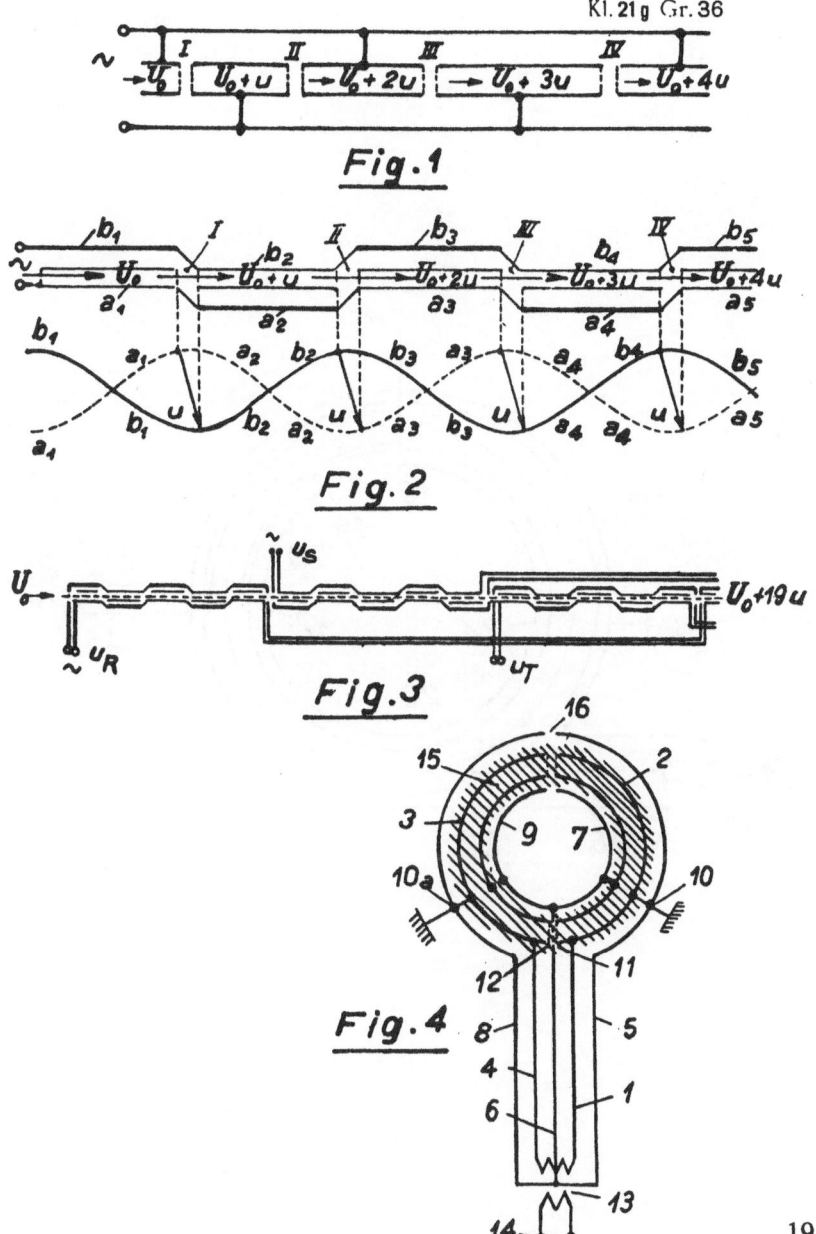

Fig.1

Fig.2

Fig.3

Fig.4

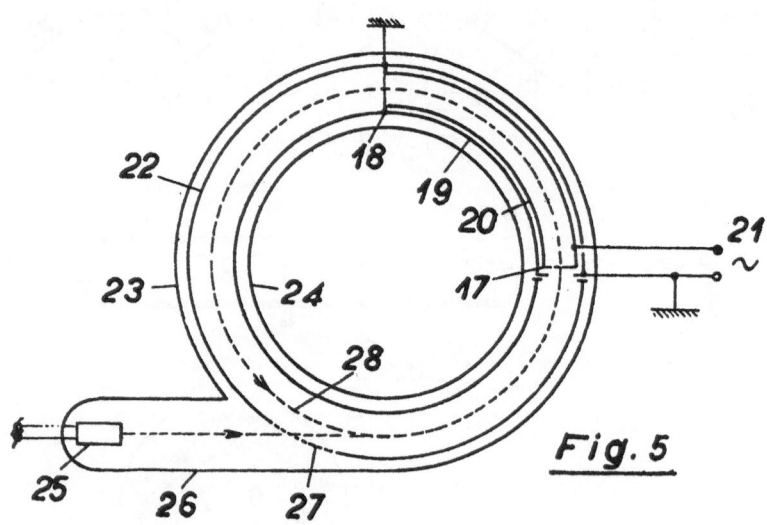

Fig. 5

Index

194

Fermilab 128
Finzi, L. 37
Fischer AG 107
Fischer, G. E. 38
Fischer, Mr. 113
Fjeld, J. 12
Flegler, Dr. 26-7, 29, 41
focusing 50, 80, 123
    strong 80, 100, 123, 167, 170
    weak 123
Ford, J. 140, 171
'Fra Fysikkens Verden' 58, 155
Frascati (Laboratories) 6, 89, 93,
    117, 146, 168, 175, 178
Frich, Ö. R. 12
Fritz-Niggli, Prof. H. 131
Fry, D. W. 122, 166
Fuhlsbüttel 4, 74, 88, 91, 161

Gaede, W. 17-8, 25, 83, 92, 152
Gamper, D. 105, 113
Gans, R. 66, 76, 162, 174
Geiger, H. 14, 75
Geist, F. 71
General Electric 59, 68, 110, 116,
    156, 172
Gentner, W. 85, 122, 125, 160,
    166-7
Gerber, Herr 113
Gerke, H. 178
Gerlach 85
German Aviation Ministry 71, 76-
    7, 86, 158, 160, 162
German Research Council 76, 158
Gestapo 3, 85, 91, 161
Giel, R. 3
'gigator' 163, 176, 178
Gittelman, B. J. 90
Glaus, B. 2
glucose, 2-deoxy-D (2-DG) 138-9
Gonella, L. 117, 120, 168, 171
Goebbels, J. 89

Gooch, P. C. 135
Goward, F. K. 100, 119, 122-4,
    164, 166-7, 171
Gräf, W. 113
Greinacher, H. 51, 171
Grini (concentration camp) 95
Groß-Ostheim 77, 160
Guddal, M. 2
Gund, K. 66, 68, 76, 79, 84-5, 91,
    110, 157, 160, 164-5, 171-2

Hafstad, L. R. 51, 153, 170
Hagemann, Edith 3
Haggard, R. 12
Haissinski, J. 94, 171
Hansen, A. 9, 19
Hansen, E. 55, 57
Hardt, Dr. W. 127
Hartmann, H. (DESY) 38
Hartmann, H. (BBC) 102, 164
Haxel, O. 88
Haug, Dir. 52
Heessel, G. 3
Heisenberg, W 85, 161
HERA 92, 127-9, 147-8, 175
Hermann, A. 121, 171
Hitler, A. 51, 65, 71, 74, 88, 154,
    162
Hollnack, (Theodor?) 71, 74, 77,
    86, 158-9, 162, 164
Houtermans, F. G. 158
Howard-Flanders, P. 136
Hylleraas, E. 58, 96, 171

Ilebu-prison 95, 162
induction accelerator 59, 66, 165,
    172, 176
Inselspital Berne 106, 108-9, 166,
    168
ionisation 132-5, 170
ions 34-5

195

relays 5, 48-55, 57, 95, 143-4, 146, 153-5
relativity (special) 12, 18, 21, 42
Rennäs, Dr. 139
repopulation (of cells) 136, 141
Rheotron 66, 76, 158, 161-2
Richter, R. 16-17
Richter, B. 92
Rogowski, W. 26-7, 29, 32, 37, 41, 46, 48, 152-3
Rotheim, Herr 19
Rüdenberg, R. 48, 67-8, 152, 154, 174
Rudolf-Virchow-Hospital 136
Rutherford, E. 9, 12, 14, 23, 46, 64-5, 81, 151, 153
RWTH, see Aachen

Salvini, G. 117
Sand, H. 174
SAS 11
SC (CERN) 42, 117
Scherrer, P. 101, 104, 164
Schiebold, Dr. 75-6, 85, 159
Schinz, R. 104, 108, 130-1, 176
Schleiermacher, Prof. 17, 24, 152
Schmellenmeier, H. 66, 76, 158, 161-2
Schmelzer, Chr. 99, 122, 126, 166
Schumacher, W. 136-40, 174
Schumann, G. 85-8, 161-2, 165, 172
Schwarz, W 3
Segrè, E. 1, 174
Seib, K. 3
Seibert, J. 3
Seifert, E. 70
Seifert, R. 70, 74, 86-7, 158-9
Sempert, M. 113, 174
Serber, R. 59, 66, 110, 156, 172
Siemens 157, 167

Siemens-Reiniger (Erlangen) 68, 79, 84-5, 91, 110-2, 160-1, 165-6
Siemens-Schuckert (Berlin) 48, 55, 65, 67-8, 152, 154-8, 173-4
SLAC 44, 90, 115, 140, 171
Slepian, J. 62-64, 66, 68, 110, 151-2, 155, 174
Sloan, D. 39, 43, 153, 172
Smekal, Prof. 88
Smidt, Dr. 64
Smithsonian Inst. (Museum) 36, 79
Snyder, H. 100, 167, 170
Solberg, Dir. 57
Sommerfeld, A. 27, 155, 158, 171
Sommerfeld, E. 27, 51, 69, 80, 82, 145, 158, 164
Spannhake, Prof. 17
stability (of the orbits) 6, 39, 65-8, 73, 83, 153-5
storage rings 4, 6, 44, 82-4, 92-4, 127-9, 143-6, 148, 159, 167-8, 173, 174, 178-9
for electrons 82, 90, 92-4, 128, 148, 168, 175
for protons 92, 94, 128, 171
SSC (reports) 169, 171
strong focusing, see focusing
Stavanger, G. von 19
Steen, Dr. 111, 166
Steenbeck, M. 48, 65-9, 79, 84, 110, 152, 154-8, 167, 174
Ström, K. 11
Stüben, A. 3
Styff, Ing. 57, 155
Suess, H. 88
Sutton, Chr. 1, 170
Swinne, E. 3, 76, 174